Math Fundamentals

Ten Pups

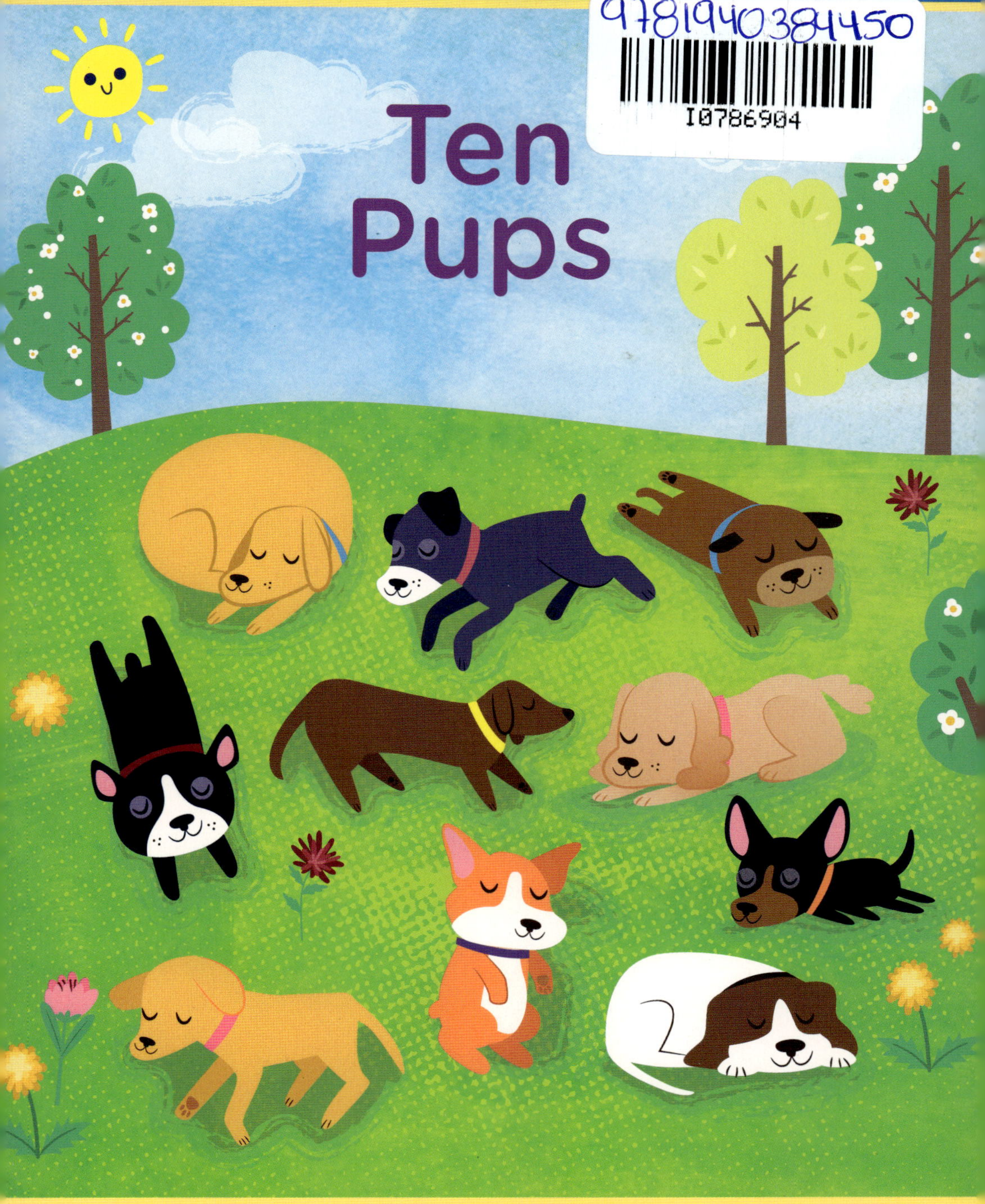

Written by Cynthia Stierle
Illustrated by Doreen Mulryan

Ten Pups

Written by Cynthia Stierle
Illustrated by Doreen Mulryan

Design by Bethany Schadel

Published by Sandviks HOP, Inc.

Printed in China.

18015

How many pups are ready for some fun?

$0 + 10 = 10$

None are awake!
Ten pups nap in the sun.

How many pups are still in their beds?

One pup is up!

1 + 9 = 10

Nine rest
their heads.

How many pups are ready to play?

$2 + 8 = 10$

Two pups wiggle, while eight start their day.

How many pups see something flutter around?

3 + 7 = 10

Three pups look up,
while seven romp
on the ground.

How many pups decide to give chase?

4 + 6 = 10

Four scamper off,
while six stay in place.

How many pups must stop their run?

5 + 5 = 10

Five pups stay still,
while five join the fun.

How many pups want to dig, run, or leap?

$10 + 0 = 10$

Ten pups on the go,
and none are asleep!

Ten happy pups are
tired once more.

They curl up together
and soon start to snore.

HOOKED ON MATH

Steps 1, 2, & 3

18015 (05/22

Math Fundamentals

The Lemonade Stand

Written by Chris Sawyer
Illustrated by Nadeem Zaidi

The Lemonade Stand

Written by Chris Sawyer
Illustrated by Nadeem Zaidi

Use *The Lemonade Stand* sticker sheet and a pencil with this book.

Design by Bethany Schadel

Published by Sandviks HOP, Inc.

Printed in China.

18016

Tali sees a big lemon tree. She gets an idea. Today she will make lemonade!

Tali picks 4 lemons. She needs 3 more.

Find 3 lemons and add them to the basket.

How many lemons are in the basket now?

$$4 + 3 = \underline{\qquad\qquad}$$

At home, Tali fills a pitcher with water. She squeezes lemon juice into the water.

Tali adds 2 lemon slices. She needs 5 more.

Find 5 lemon slices and add them to the pitcher.

How many slices are in the pitcher now?

$2 + 5 =$ ______

Tali takes a sip. *Eeek!* It is sour!

"I'll ask the bees for some honey," she says.

The bees are happy to share. But first, they need to visit the flowers to make more honey.

Find the bees and stick them onto the flowers.

How many flowers in all? ☐

How many bees in all? + ☐

How many bees and flowers all together? ☐

Now the lemonade is perfect!

Next, Tali builds a lemonade stand. Her friend Chow helps her.

They hammer 7 boards into place. They need to add 2 more.

Find 2 planks of wood and add them to the stand.

How many planks does the stand have now?

$7 + 2 =$ ______

Next, Batu drives up to the stand.

Tali gives him some lemonade.

"Thanks!" says Batu. "I am planting trees today. Would you like me to plant one for you?"

"Oh, yes!" Tali says. "There are no trees in that spot. A tree would be nice."

Find 1 tree and place it in the hole.

How many trees does Batu plant for Tali?

$0 + 1 =$ ______

"Look!" Batu shouts. "A parade!"

Kazoo and his meerkat band reach the lemonade stand.

Tali gives them all lemonade.

"Thank you," Kazoo tells Tali. "Now we can keep marching. We will line up from 1 to 10."

Find the missing meerkats and add them in the correct order from 1 to 10.

Kazoo looks at Tali and Batu. "We could use 2 more in our parade. Would you both like to join our band of 10?"

Tali and Batu say yes!

The band has 10 players.

Add Tali and Batu to the scene so they can join the marching band.

Now how many players are there?

$$10 + 2 = \_\_\_\_\_\_$$

The lemonade is finished, but the fun will keep marching on!

HOOKED ON MATH

Steps 1, 2, & 3

18016 (05/22

Math Fundamentals

MATH

The Collector

Written by Chris Sawyer
Design by Bethany Schadel

The Collector

Written by Chris Sawyer

Design by Bethany Schadel

Published by Sandviks HOP, Inc.

Printed in China.

18030

My house may seem like quite a mess,
but it looks just fine to me.
That's because I find a way
to sort mathematically.

There are 60 socks in my collection.
I like the striped ones best.
With 40 of those to warm my toes,
my dog can wear the rest.

How many socks don't have stripes?

$$\begin{array}{r} 60 \\ -\ 40 \\ \hline 20 \end{array}$$

Bonus: Can you find a sock with cow spots?

I like to wear sunglasses.

I had 47 pairs.

Then I went and got 12 more

to dress my teddy bears.

How many pairs of sunglasses in all?

$$\begin{array}{r} 47 \\ +\ 12 \\ \hline 59 \end{array}$$

Bonus:
Can you find the heart-shaped sunglasses?

I have 81 bricks that click.

I can stack them short or tall.

I just found 16 yellow ones

so I can build a wall!

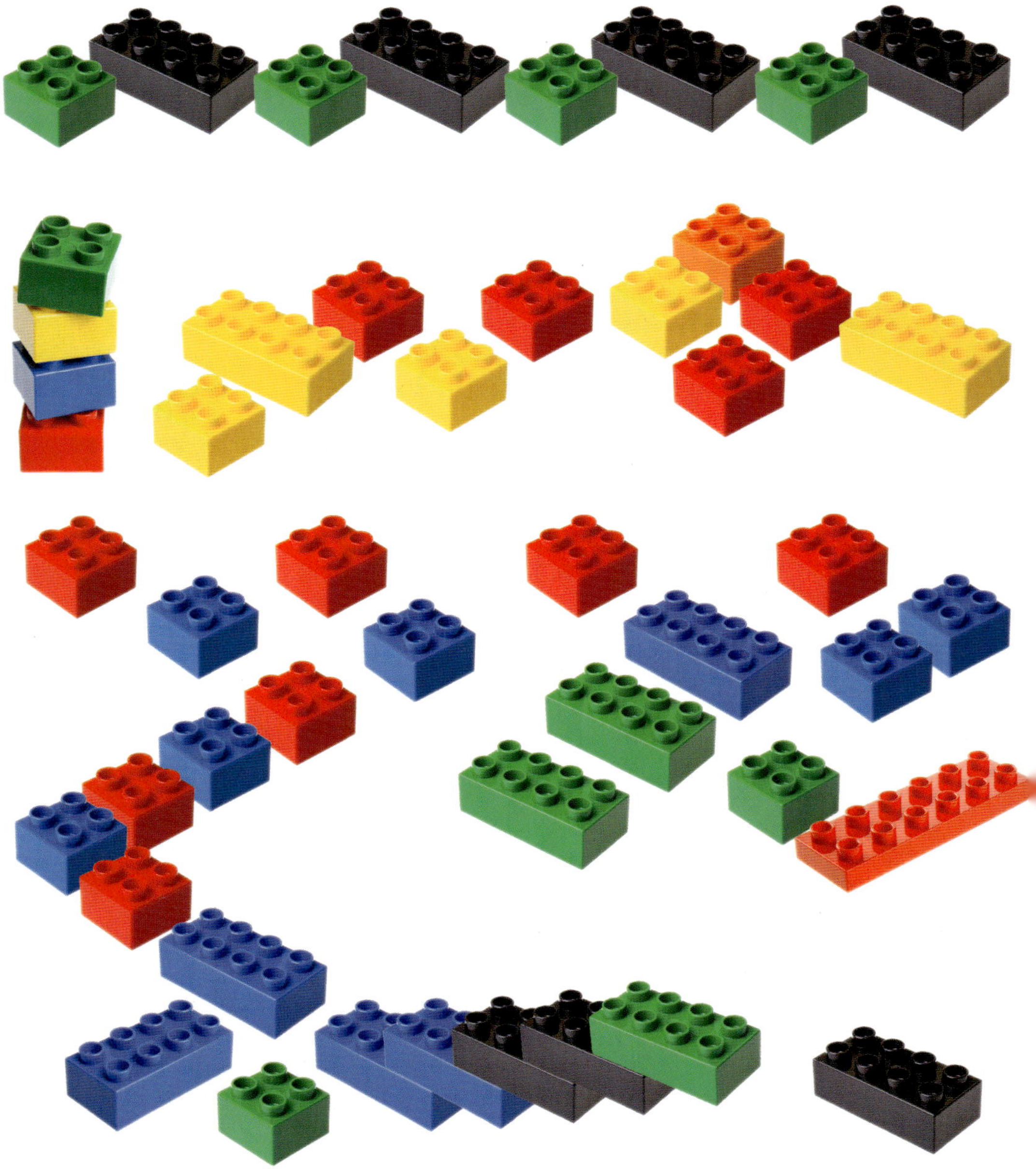

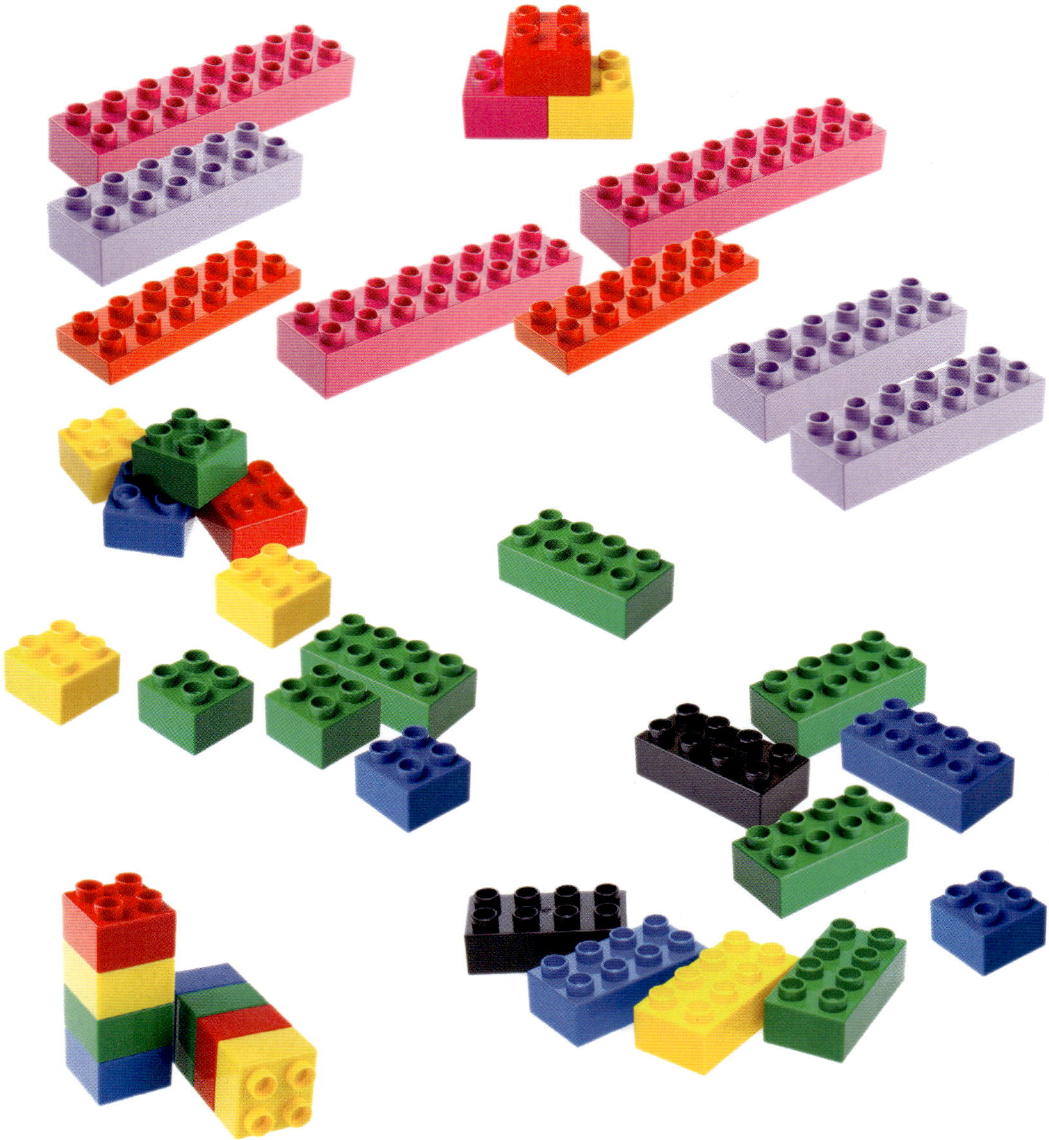

How many building bricks in all?

$$
\begin{array}{r} 81 \\ +\ 16 \\ \hline 97 \end{array}
$$

Bonus:
Can you find the orange brick?

With 68 sharp pencils,

I made a birthday card.

But 17 snapped off their tips.

I guess I pressed too hard.

How many sharpened pencils are left?

$$\begin{array}{r} 68 \\ -\ 17 \\ \hline 51 \end{array}$$

Bonus:
Can you find the white pencil?

Look at my shiny stones!

I have 79 beauties so far.

The 22 green ones are my favorites.

The rest I save in a jar.

How many stones are *not* green?

$$\begin{array}{r} 79 \\ -\ 22 \\ \hline 57 \end{array}$$

Bonus:
Can you find the gold nugget?

You'll never guess what I collect,
but I think it's pretty nifty:
Sneakers of all shapes and sizes—
I own 23 plus 50!

How many sneakers in all?

$$\begin{array}{r} 23 \\ +\,50 \\ \hline 73 \end{array}$$

Bonus:
Can you find the orange shoelaces?

I love collecting seashells.

I have exactly 57.

But starfish are not an easy find.

Of those, I have only 11.

How many seashells and starfish in all?

$$\begin{array}{r} 57 \\ +\ 11 \\ \hline 68 \end{array}$$

Bonus: Can you find the hermit crab?

It's been so much fun to share with you.

I hope you liked my stuff.

You can go back and look again

if you can't get enough!

Can you spot a:

sock with a
pink flamingo?

red and blue sneaker?

tower of
five building
bricks?

white
starfish?

broken
orange
pencil?

teddy bear with
a baseball cap?

stone with black
and purple flecks?

HOOKED ON MATH

Steps 10 & 11

18030 (07/22

Math Fundamentals

The Seesaw

Written by Donna Giachetti
Illustrated by Nadeem Zaidi

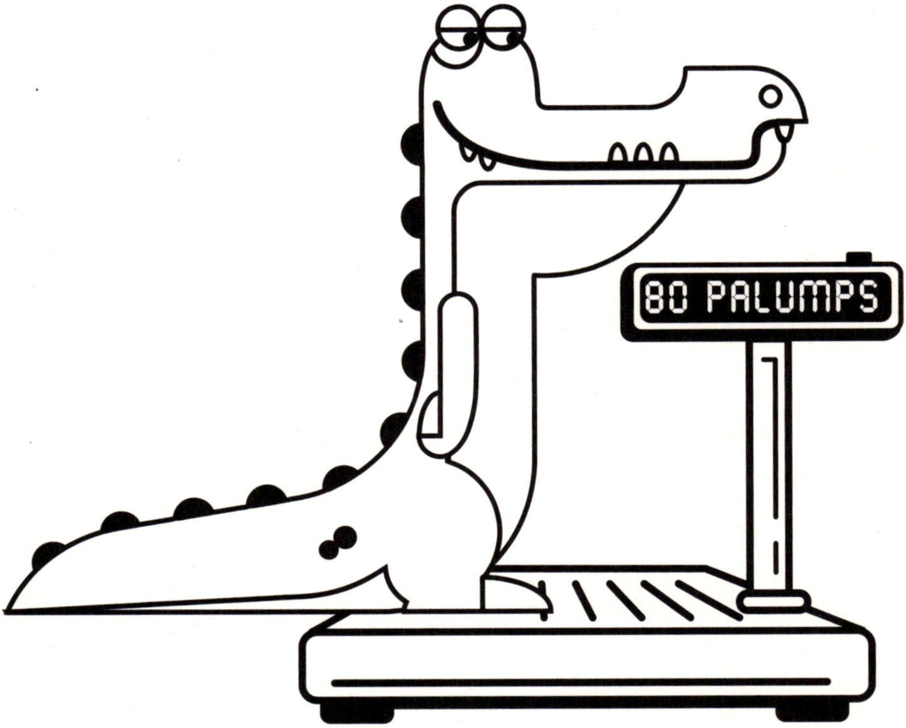

In this story, the pals measure weight in the fictional unit of *palumps*.

The Seesaw

Written by Donna Giachetti
Illustrated by Nadeem Zaidi

Design by Bethany Schadel

Published by Sandviks HOP, Inc.

Printed in China.

18029

Chow has built a new seesaw for the playground. Time to try it out!

"Hello?" he calls. "Anybody?"

All alone, Chow settles himself on one side of the seesaw. BUMP! He drops all the way down to the ground.

80 PALUMPS > 0 PALUMPS

"Not as much fun as I hoped," he mumbles.

Happily, some meerkats have shown up to play. Meerkats are always curious about new things.

Kip asks, “Can we try it?”

Kip and Rosie climb on and balance perfectly.

“Wow!” Chow cheers. “You guys must be twins.”

10 PALUMPS = 10 PALUMPS

Zola, Gabby, and Tali arrive. "Can we have a turn?"

"Sure!" says Kip. He and Rosie jump off.

Zola, Monty, and Miles sit on one side.

Tali and Gabby sit on the other.

Zola's group goes way up in the air. "Woo-hoo!" Zola calls. "I can see everything from up here!"

55 PALUMPS > 25 PALUMPS

Now Kazoo comes along. “You need more weight on your side,” he tells Zola.

He climbs onto the seesaw, but it’s still not enough to budge Tali and Gabby.

So he calls the rest of the meerkats over.

Suddenly, Tali and Gabby are up in the air.

“See?” Kazoo says. “That’s how it’s done.”

55 PALUMPS < 85 PALUMPS

"I'll help you balance," Chow says.

He wiggles his way onto Tali and Gabby's side of the seesaw.

"Make room!" he grunts.

It's a tight fit, but he manages.

Kazoo and his pals go up, up, up!

135 PALUMPS > 85 PALUMPS

Kazoo slips off the seesaw. “I’ll be right back.”

He looks different when he returns.

Zola eyes him curiously. What is he carrying?

Kazoo climbs back onto the seesaw.

Ta-da! His side is now balanced with Chow’s side.

135 PALUMPS = 135 PALUMPS

Along comes Batu. “Me, me!” he cries. “I love to seesaw!”

He joins Kazoo’s side. The seesaw creaks and groans. Chow, Tali, and Gabby shoot way up.

“We’re not balanced!” Tali cries out. “Let us down!”

Batu looks sad. “But I want to play.”

135 PALUMPS < 335 PALUMPS

"Let's start over," Zola suggests.

All the pals climb down off the seesaw.

0 PALUMPS = 0 PALUMPS

Kazoo takes off his backpack. “I can’t carry any more rocks.”

Batu says, “I have a plan.”

He arranges his pals on one side of the seesaw.

“That’s just about everybody,” Zola says. “Except for you, Batu.”

0 PALUMPS < 200 PALUMPS

Batu needs help climbing onto the empty side of the seesaw.

Meri and Kat hold the ladder steady.

"I'm coming," Batu huffs. "Just need to . . . get my leg . . . on the other side."

The seesaw is balanced once again.

Batu grins. "I feel like I'm flying!"

The pals cheer!

“Thanks, Batu,” says Chow. “You didn’t let us down.”

200 PALUMPS = 200 PALUMPS

A tiny butterfly lands on Gabby's beak with a gentle flutter.

Suddenly, Batu is launched into the air!

The pals are amazed.

Chow says, "Gee, I never knew a butterfly could make a rhino fly!"

Create your own story with the Hooked on Math friends. How will you stack them up on the seesaw? Try different ways of making their weights balance. For example: Tali + Kazoo + Rosie = Chow.

BATU	**200 palumps**
CHOW	**80 palumps**
TALI	**50 palumps**
KAZOO	**20 palumps**
KAZOO WITH ROCKS	**70 palumps**
MONTY, MERI, KIP, KAT, MILES, ROSIE	**10 palumps each**
GABBY	**5 palumps**
ZOLA	**5 palumps**

HOOKED ON MATH

Steps 10 & 11

hookedonmath.com

18029 (07/22

Math Fundamentals

100 Fingers for 100 Days

Written by Cynthia Stierle
Illustrated by Aram Song

100 Fingers for 100 Days

Written by Cynthia Stierle
Illustrated by Aram Song

Design by Bethany Schadel

HOOKED ON MATH

Published by Sandviks HOP, Inc.

Printed in China.

18026

Miss Luff shares her idea.
The class thinks that it's cool.
They'll make a special poster
to mark 100 days of school.

She tells the kids to grab their smocks,
as she hangs paper on the wall.
Once they add their fingerprints,
they'll have 100 prints in all!

Miss Luff chooses orange paint.
She dips her hands and then
her fingers press on paper.
Next, she writes *5* and *10*.

1	2	3	4	5	6	7	8	9	10
11	12	13	14	15	16	17	18	19	20
21	22	23	24	25	26	27	28	29	30
31	32	33	34	35	36	37	38	39	40
41	42	43	44	45	46	47	48	49	50
51	52	53	54	55	56	57	58	59	60
61	62	63	64	65	66	67	68	69	70
71	72	73	74	75	76	77	78	79	80
81	82	83	84	85	86	87	88	89	90
91	92	93	94	95	96	97	98	99	100

Aaron offers to go next.
He lets out a great big roar!
He uses green, like a T-Rex,
his favorite dinosaur.

1	2	3	4	5	6	7	8	9	10
11	12	13	14	15	16	17	18	19	20
21	22	23	24	25	26	27	28	29	30
31	32	33	34	35	36	37	38	39	40
41	42	43	44	45	46	47	48	49	50
51	52	53	54	55	56	57	58	59	60
61	62	63	64	65	66	67	68	69	70
71	72	73	74	75	76	77	78	79	80
81	82	83	84	85	86	87	88	89	90
91	92	93	94	95	96	97	98	99	100

Lucas wants the purple paint.
He almost spills it in a rush.
He likes painting with his hands.
It's much better than a brush!

1	2	3	4	5	6	7	8	9	10
11	12	13	14	15	16	17	18	19	20
21	22	23	24	25	26	27	28	29	30
31	32	33	34	35	36	37	38	39	40
41	42	43	44	45	46	47	48	49	50
51	52	53	54	55	56	57	58	59	60
61	62	63	64	65	66	67	68	69	70
71	72	73	74	75	76	77	78	79	80
81	82	83	84	85	86	87	88	89	90
91	92	93	94	95	96	97	98	99	100

Quiet Lulu chooses teal.
She thinks that art is fun.
She adds ten more fingerprints,
then cleans up when she's done.

1	2	3	4	5	6	7	8	9	10
11	12	13	14	15	16	17	18	19	20
21	22	23	24	25	26	27	28	29	30
31	32	33	34	35	36	37	38	39	40
41	42	43	44	45	46	47	48	49	50
51	52	53	54	55	56	57	58	59	60
61	62	63	64	65	66	67	68	69	70
71	72	73	74	75	76	77	78	79	80
81	82	83	84	85	86	87	88	89	90
91	92	93	94	95	96	97	98	99	100

Mateo knows that Marisol
loves the color pink.
So what color should he choose?
"I'll pick red, I think."

Tess is happy to choose yellow,
a color like the sun.
She slaps the paper with a SMOOSH!
Oh, my! That was fun.

1	2	3	4	5	6	7	8	9	10
11	12	13	14	15	16	17	18	19	20
21	22	23	24	25	26	27	28	29	30
31	32	33	34	35	36	37	38	39	40
41	42	43	44	45	46	47	48	49	50
51	52	53	54	55	56	57	58	59	60
61	62	63	64	65	66	67	68	69	70
71	72	73	74	75	76	77	78	79	80
81	82	83	84	85	86	87	88	89	90
91	92	93	94	95	96	97	98	99	100

All this while the class hamster
is looking for something to munch.
Miss Luff hands some treats to Marisol.
"Here. It's time for Artie's lunch."

It's Sam's turn and she decides
to give light blue a try.
She inspects her work and says,
"It looks just like the sky."

1	2	3	4	5	6	7	8	9	10
11	12	13	14	15	16	17	18	19	20
21	22	23	24	25	26	27	28	29	30
31	32	33	34	35	36	37	38	39	40
41	42	43	44	45	46	47	48	49	50
51	52	53	54	55	56	57	58	59	60
61	62	63	64	65	66	67	68	69	70
71	72	73	74	75	76	77	78	79	80
81	82	83	84	85	86	87	88	89	90
91	92	93	94	95	96	97	98	99	100

"Your turn," Miss Luff tells Theodore.
He quickly chooses brown.
"Just like my dog!" he says—THWAP!—
and sets his fingers down.

1	2	3	4	5	6	7	8	9	10
11	12	13	14	15	16	17	18	19	20
21	22	23	24	25	26	27	28	29	30
31	32	33	34	35	36	37	38	39	40
41	42	43	44	45	46	47	48	49	50
51	52	53	54	55	56	57	58	59	60
61	62	63	64	65	66	67	68	69	70
71	72	73	74	75	76	77	78	79	80
81	82	83	84	85	86	87	88	89	90
91	92	93	94	95	96	97	98	99	100

Marisol is last to go.
She picks her spot with care.
She adds her pink fingerprints.
Now prints are everywhere!

1	2	3	4	5	6	7	8	9	10
11	12	13	14	15	16	17	18	19	20
21	22	23	24	25	26	27	28	29	30
31	32	33	34	35	36	37	38	39	40
41	42	43	44	45	46	47	48	49	50
51	52	53	54	55	56	57	58	59	60
61	62	63	64	65	66	67	68	69	70
71	72	73	74	75	76	77	78	79	80
81	82	83	84	85	86	87	88	89	90
91	92	93	94	95	96	97	98	99	100

But to reach 100,
they need another 10.
Since they've all had a turn,
who gets to go again?

Miss Luff says that someone
hasn't gone just yet.
She goes to the hamster cage
and lifts out the class pet.

"Artie!" the class cries at once.
Though his back paws are quite small,
with great care Miss Luff adds
his paw prints to the wall.

Now there are 100 prints!
The kids all shout "Hooray!"
for their special art creation
and for the 100th day!

1	2	3	4	5	6	7	8	9	10
11	12	13	14	15	16	17	18	19	20
21	22	23	24	25	26	27	28	29	30
31	32	33	34	35	36	37	38	39	40
41	42	43	44	45	46	47	48	49	50
51	52	53	54	55	56	57	58	59	60
61	62	63	64	65	66	67	68	69	70
71	72	73	74	75	76	77	78	79	80
81	82	83	84	85	86	87	88	89	90
91	92	93	94	95	96	97	98	99	100

1	2	3	4	5	6	7	8	9	10
11	12	13	14	15	16	17	18	19	20
21	22	23	24	25	26	27	28	29	30
31	32	33	34	35	36	37	38	39	40
41	42	43	44	45	46	47	48	49	50
51	52	53	54	55	56	57	58	59	60
61	62	63	64	65	66	67	68	69	70
71	72	73	74	75	76	77	78	79	80
81	82	83	84	85	86	87	88	89	90
91	92	93	94	95	96	97	98	99	100

Color the numbers to skip count by 5s to 100. You can use the same color for every number or a variety of colors to match the ones in the story.

HOOKED ON MATH

Steps 8 & 9

18026 (07/2

Math Fundamentals

Kazoo and the Seed

Written by Chris Sawyer
Illustrated by Nadeem Zaidi

Note to Parents

This imaginative story includes visual representations of one-digit and two-digit numbers. Numbers appear in the place value chart and as place value blocks (ones) and rods (tens). Also known as base-ten blocks, your child may have encountered these as physical manipulatives in the classroom. These important math tools help your child understand that where a digit is placed within a number relates to an amount, which is a key concept when working with numbers both big and small.

TENS	ONES
6	5

Place Value Chart

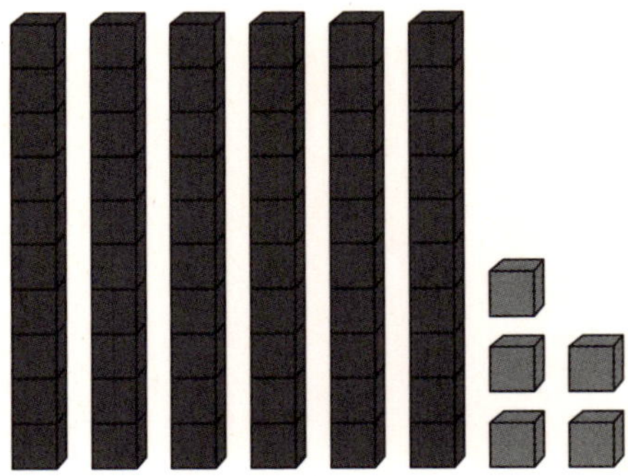

Place Value Rods and Blocks

In this story, the amount and colors of the leaves correspond to the numbers in the place value chart, as well as the ones blocks and tens rods. It's important for your child to recognize that a rod is equal to a group of ten blocks. Also, as the number of leaves on the tree grows, help your child see that skip counting by groups of ten is faster than counting by ones. (Try it and see!)

Kazoo and the Seed

Written by Chris Sawyer
Illustrated by Nadeem Zaidi

Design by Bethany Schadel

Published by Sandviks HOP, Inc.

Printed in China.

18027

Batu gasps when he looks at his calendar. "Uncle Rhino will be here in ten days! I really must tidy up."

It takes him many hours to clean his hut. He carries the clutter outside and puts it in a garbage can.

Kazoo is walking down the road when he spots something that looks like a speckled grape.

"Why, it's a seed!" he realizes. "Where did you come from?" He takes the seed home. He plants it in a pretty pot and sets it near a sunny window.

The next day, Kazoo sees a tiny sprig poking out of the soil. One shiny pink leaf has popped out, too!

"What a pretty color! I'll have to show Tali."

Then he gives his new plant a drink of water.

TENS	ONES
	1

When Tali comes to visit, she brings Kazoo some plant food.

His seedling now has three pink leaves.

"Pink!" she says. "That's my favorite color."

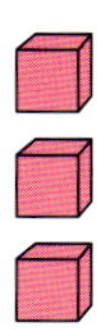

TENS	ONES
	3

Kazoo scatters a bit of plant food on the sprout. "This will help you grow even faster," he tells it.

Before he finishes speaking, four more leaves pop open.

"That was fast," Tali gasps. "We'll have to tell Zola."

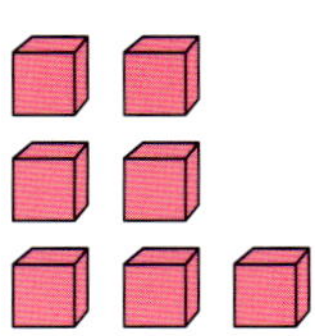

TENS	ONES
	7

When he wakes up the next day, Kazoo sees more leaves. Now there are eleven! Some have even turned from pink to purple.

He sings to his plant all day and reads to it at night before bed. “Good night, little buddy,” he tells the plant as he shuts off the light.

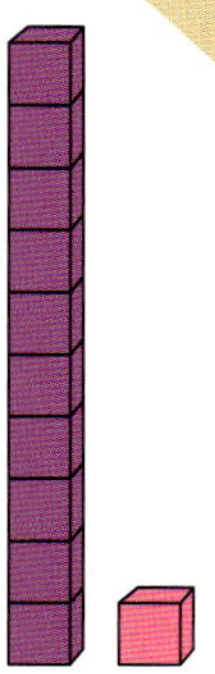

TENS	ONES
1	1

The following day Zola stops by. She knows a lot about gardening. She looks at the leaves and declares, “This plant is ready to live outdoors.”

Kazoo looks worried. “It’s not too soon?” he asks. Zola is certain.

“Okay,” Kazoo agrees. “I’ll put it outside tomorrow.”

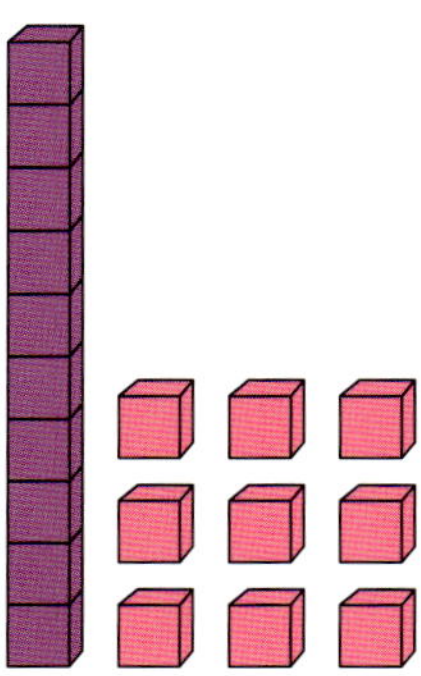

TENS	ONES
1	9

The sun is shining when Kazoo brings the pot outside the next morning.

He carries the pot from place to place, until he finally picks the perfect sunny spot.

"You're a big plant now, but you'll always be my little seed," Kazoo tells the plant.

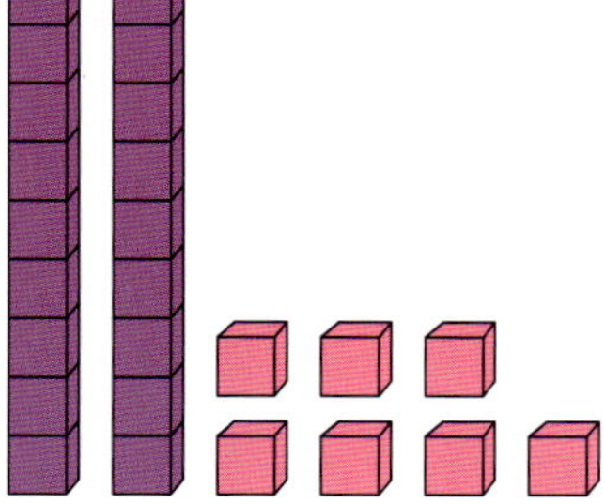

TENS	ONES
2	7

Kazoo is so proud of his little sapling. But he can't help worrying about leaving it outdoors.

"What if it's cold?" he wonders.

He brings a blanket outside, just in case.

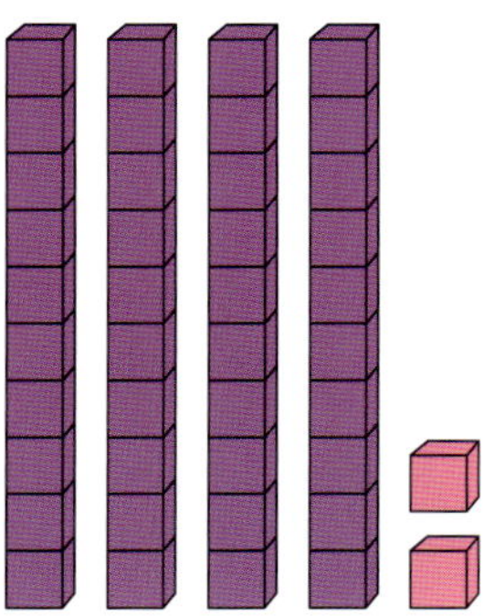

TENS	ONES
4	2

Soon Kazoo sees that Zola was right. His little shrub is happy outside!

Kazoo decides it needs a bigger pot. He calls Chow. "Can you build a planter for my tree?"

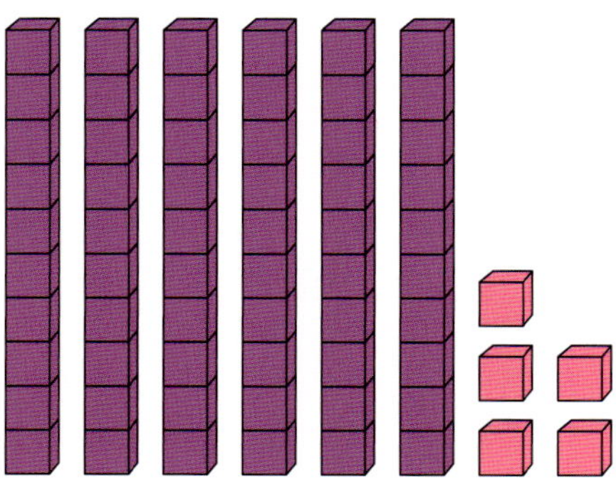

TENS	ONES
6	5

It takes all day, but the planter is finally built.

Together, Chow and Kazoo move the tree into its new home.

"Just in time," Kazoo says. "Who knows how big it will be tomorrow?"

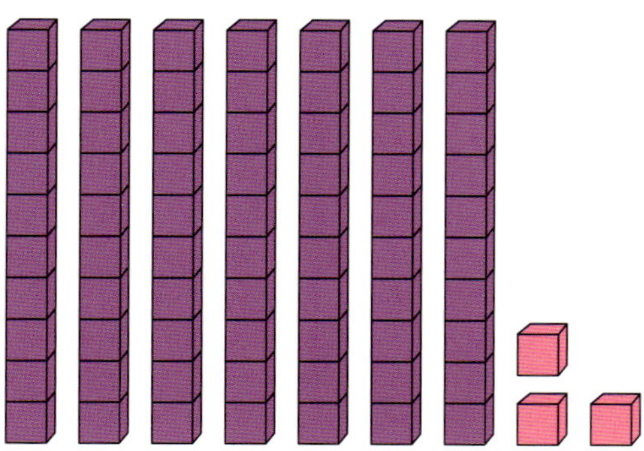

TENS	ONES
7	3

Kazoo invites all his pals to see the tree—including Batu's Uncle Rhino.

"This is a fine tree," Tali tells Kazoo. Zola and Gabby agree. "We have never seen one like this."

Uncle Rhino pipes up. "It's a ten tree."

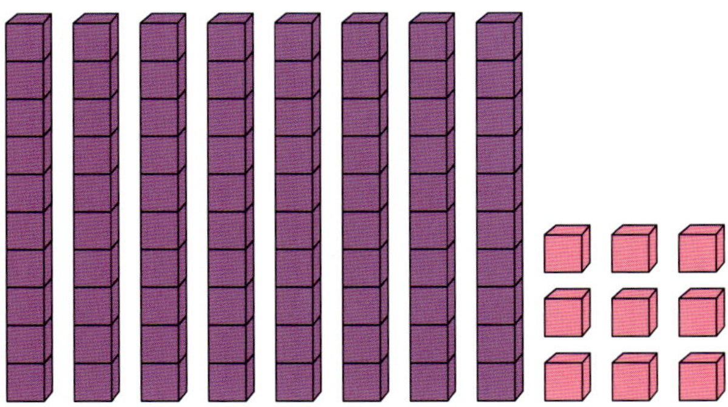

TENS	ONES
8	9

“A what?” Kazoo asks.

“A ten tree. It grows in ten days,” Uncle Rhino explains. “And I think it’s almost ready.” He reaches for a branch and shakes it gently.

Just then, all the leaves change colors.

The friends stare at the tree in wonder.

“When it’s grown, it has 100 leaves in ten different colors,” Uncle Rhino says. He looks at Kazoo. “It’s really quite tricky to grow. How did you do it?”

Kazoo smiles. “I just gave it lots of love and **TEN**der care.”

HUNDREDS	TENS	ONES
1	0	0

HOOKED ON MATH

Steps 8 & 9

18027 (07/22

Math Fundamentals

The Carnival

Written by Donna Giachetti
Illustrated by Aram Song

The Carnival

Written by Donna Giachetti
Illustrated by Aram Song

Design by Bethany Schadel

Published by Sandviks HOP, Inc.

Printed in China.

18022

Arnie and Zack want to join their friends at the carnival.

It costs 20 cents for 2 tickets.

Arnie has 14 cents.

Zack has 6 cents.

Can Arnie and Zack join their friends?

Yes!

14 + 6 = 20

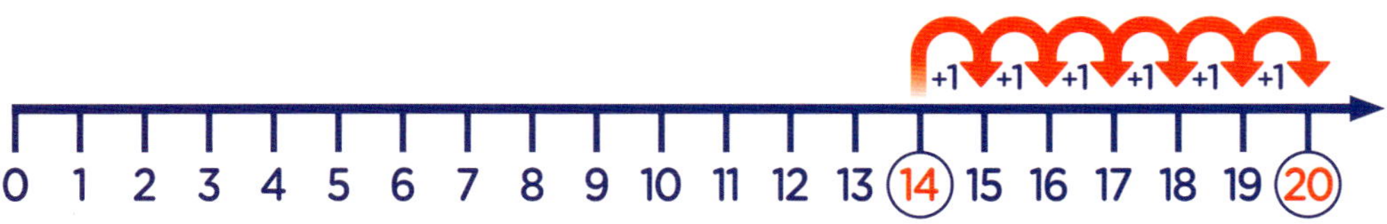

Arnie and Zack see 6 friends on the Ferris wheel.

"Let's take a ride!" Arnie says.

The 2 pals climb aboard.

How many friends are riding now?

$6 + 2 = 8$

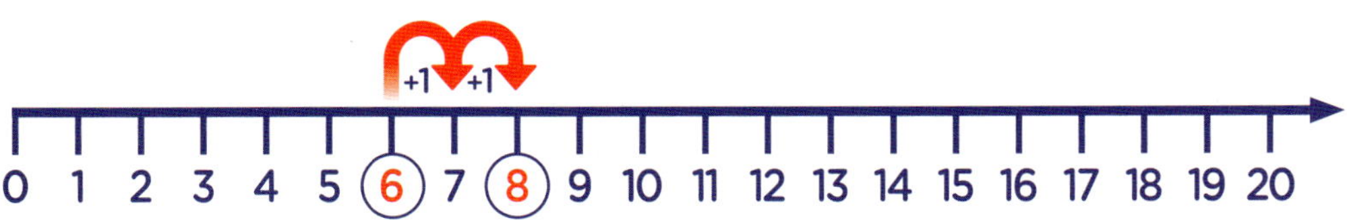

"What should we do now?" Zack asks.

His pal Bingo sees a tossing game. "This looks like fun."

Who can throw the most balls into the holes?

Bingo gets 12 balls in.

Zack gets 3 balls in.

Bingo wins a balloon!

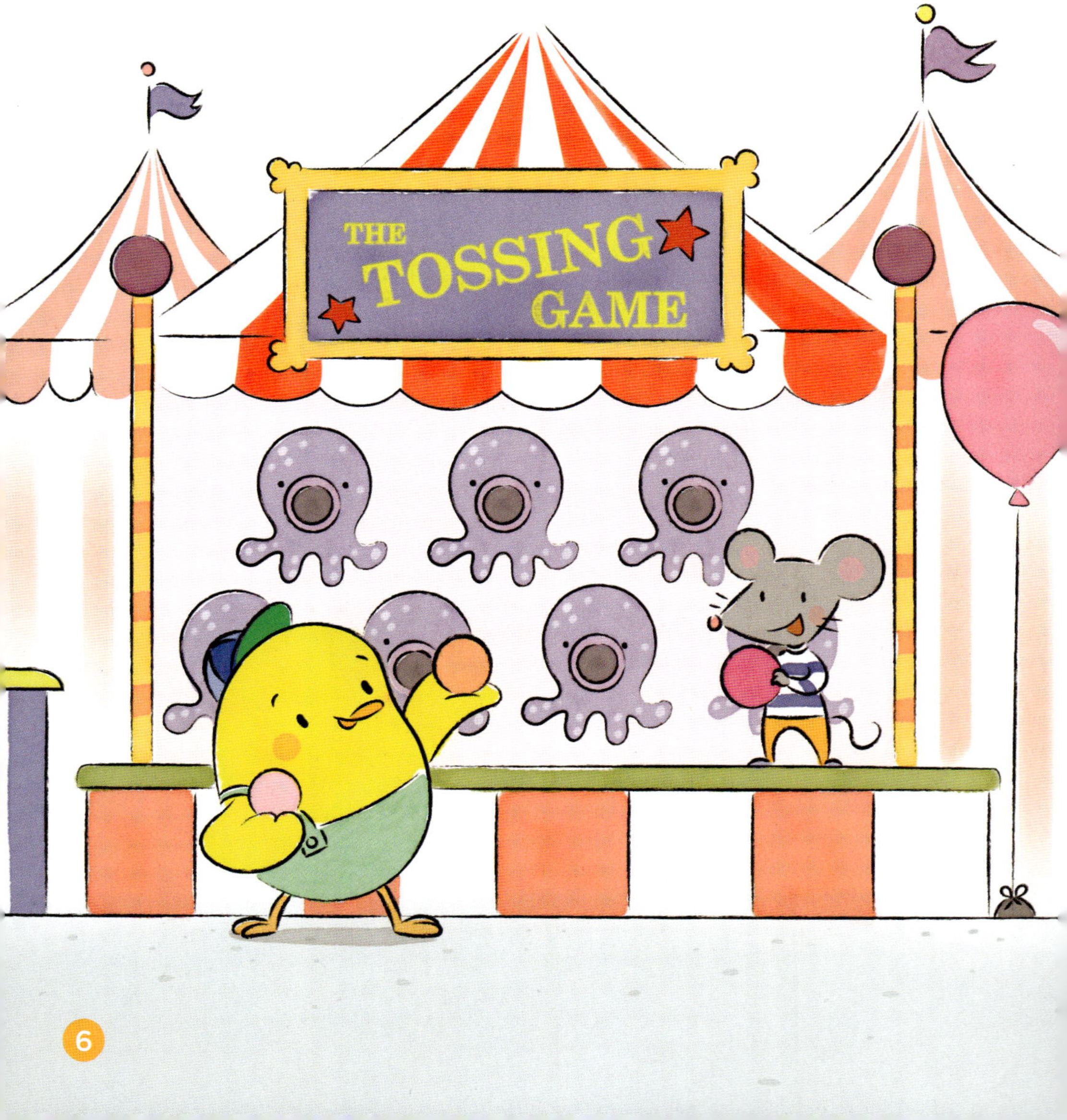

How many balls do they get in all together?

12 + 3 = 15

Zack sees Mimi run into the fun house.

She sees 10 crazy copies of herself in the mirrors, but there is only 1 real Mimi.

Zack peeks in. "Wow!" he says. "I see Mimis everywhere!"

How many Mimis does Zack see in all?

10 + 1 = 11

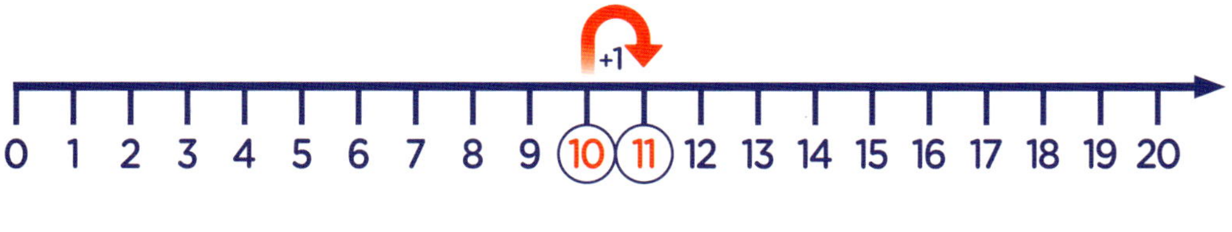

Now Zack really wants to win a game.

“This looks easy,” Hank says.

He spins twice. He gets a 5 and a 9.

“Let me try,” Zack begs.

What was the sum of Hank’s spins?

$5 + 9 = 14$

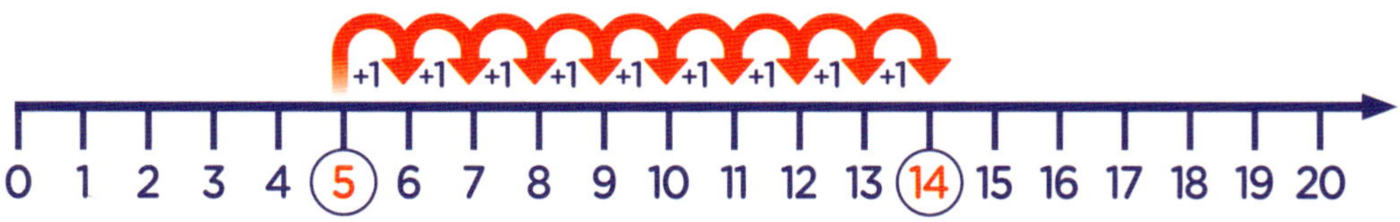

Zack spins twice. He gets a 9 and a 5.

"This game is harder than I thought," Hank says.

What was the sum of Zack's spins?

Which number line is faster to count?

9 + 5 = 14

Meanwhile, 3 pals are at the top of the slide.

Off they go—down, down, down!

Bingo, Ellie, and Mimi are watching below.

“Wait for us!” Bingo calls.

If the 3 pals below join the 3 pals above, how many are going down the slide in all?

3 + 3 = 6

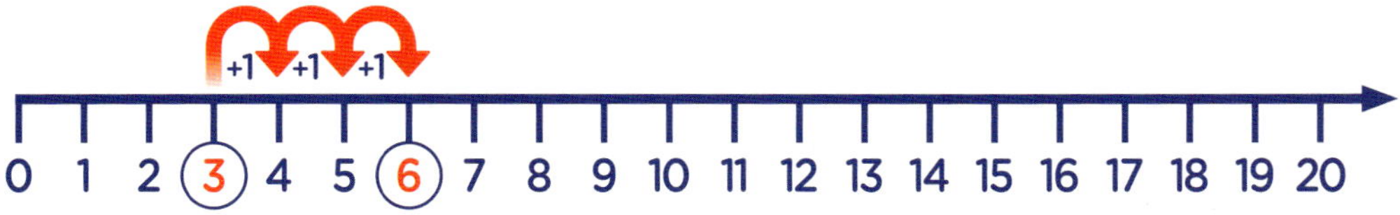

The day is almost over but Zack still wants to win a prize.

The carnival cow says, “Guess the right number and win this jar of jelly beans!”

Zack steps near the jar and points to his nose. “I can count them with this.”

Sniff! Sniff!

"That's 13," he says. *Sniffff!* "And another 7!"

What number does Zack guess?

13 + 7 = 20

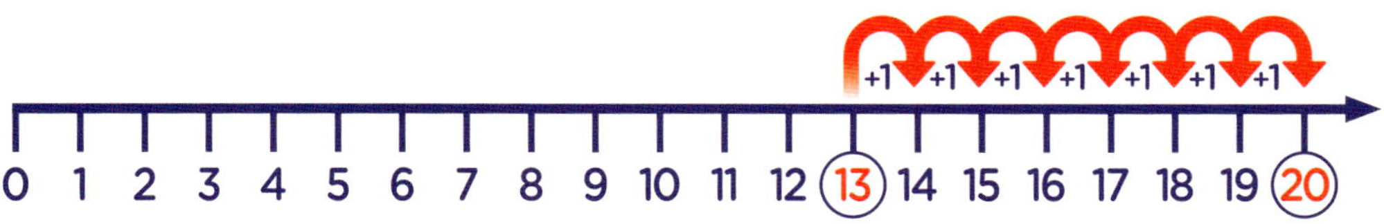

It's time to go home now.

"What did you like the most at the carnival?" Arnie asks Zack.

Zack smiles and holds up his jar.

"Sharing my prize with all my best friends!"

HOOKED ON MATH

Steps 6 & 7

Math Fundamentals

A Day at the Beach

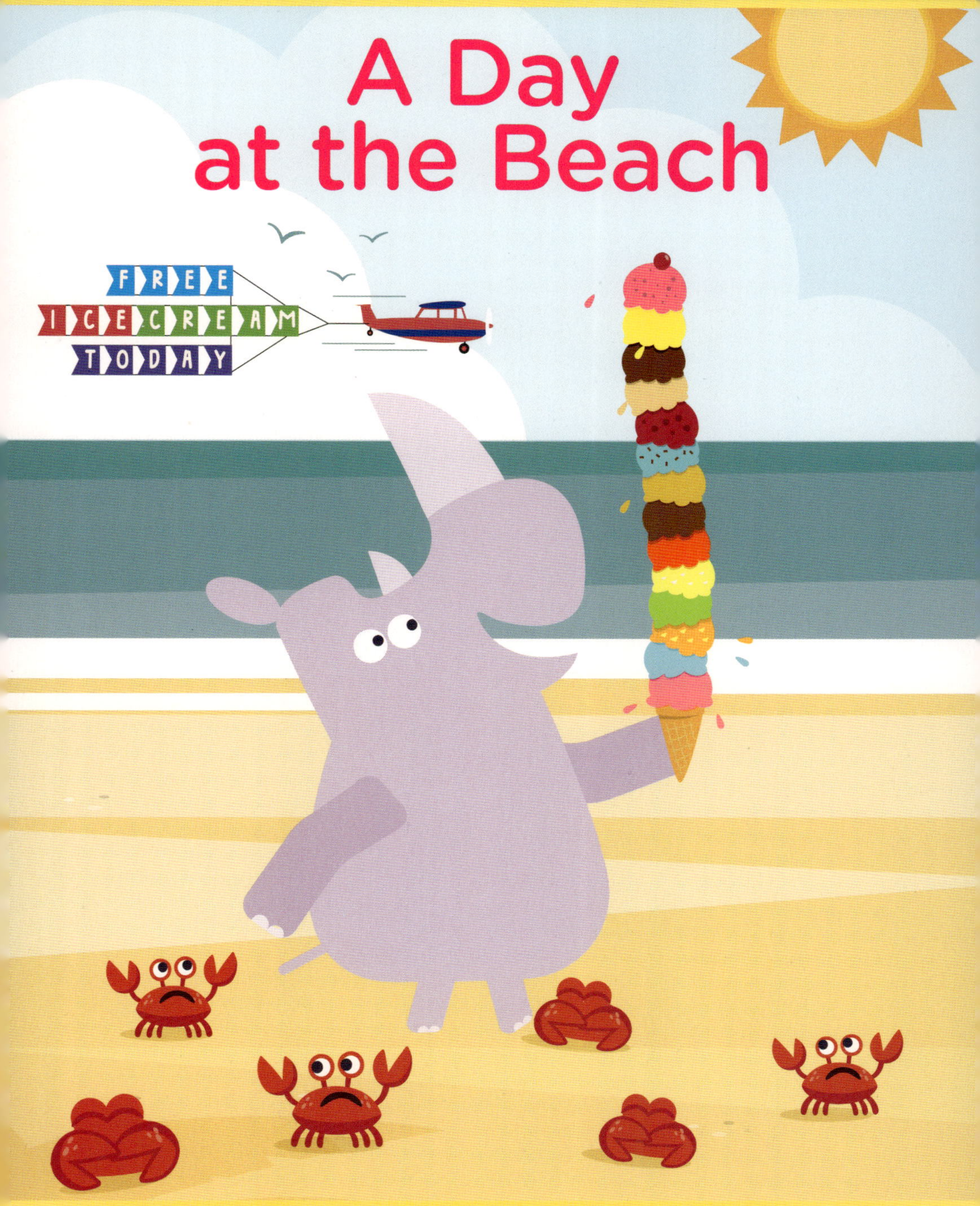

Written by Cynthia Stierle
Illustrated by Nadeem Zaidi

Note to Parents

This book is designed to be used with a pencil and the sticker sheet marked *A Day at the Beach*.

This unique sticker book is a creative way for your child to have fun with math. The story demonstrates part-part-whole reasoning. Sometimes called *number bonds*, part-part-whole problems are graphic representations that show the relationships between numbers. This helps learners recognize that whole numbers can be split into parts.

As you read together, allow your child to finish the story by placing the correct number stickers into the part-part-whole boxes on each page. Then, help them solve the problem and write the correct answer to each subtraction sentence.

A Day at the Beach

Written by Cynthia Stierle
Illustrated by Nadeem Zaidi

Have a pencil and the sticker sheet for *A Day at the Beach* handy.

Design by Bethany Schadel

HOOKED ON MATH

Published by Sandviks HOP, Inc.

Printed in China.

18023

It's a fine day for the beach. As the pals arrive, a plane zooms across the sky. It is pulling 17 letter flags.

Batu reads the words. "Free ice cream today? Wow!" He dashes off.

A strong gust of wind comes along.

Whoosh! The wind blows away 4 of the flags.

How many flags are left?

WHOLE

PART PART

$17 - 4 =$ _____

Place the number stickers onto the correct spots in the part-part-whole model. Write the answer to the question.

Zola and Tali go for a stroll along the shore.

They spot 12 shells. Zola collects 5 blue ones. The rest of the shells are red.

Tali picks one up.

The red shells are crabs!

"Put me down!" the crab tells Tali.

"So sorry!" Tali says.

How many crabs are there?

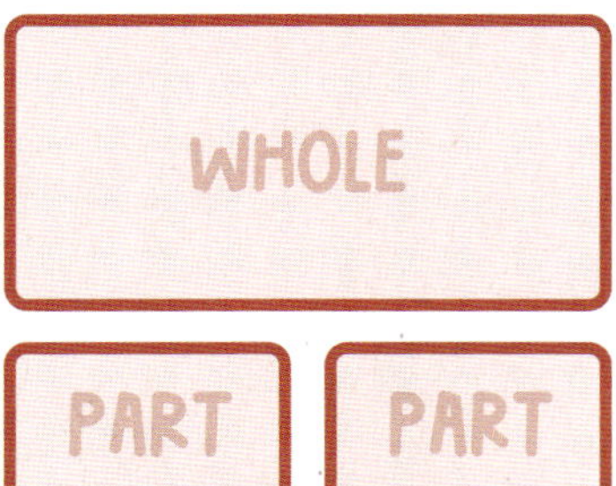

$12 - 5 =$ ______

Place the number stickers onto the correct spots in the part-part-whole model. Write the answer to the question.

Just then, Batu returns carrying an ice cream cone with 14 scoops.

It's so tall that Batu doesn't see the crabs. He trips over them. Batu drops 11 scoops of ice cream. *Splat!*

"I'm very sorry," Batu tells the crab.

"Banana? Yuck!" the crab grumbles.
"I like mint chip."

How many scoops are left on the cone?

WHOLE

PART PART

$$14 - 11 = \_\_\_\_\_$$

Place the number stickers onto the correct spots in the part-part-whole model. Write the answer to the question.

Batu goes back to the ice cream stand.

The crabs scuttle off into the water. Chow wants to join them. “Let’s go surfing,” he says.

The 4 pals paddle into the water. They each have a surfboard. Soon a big wave comes.

“Watch out!” Zola shouts, as 2 of them fall into the water. *Splash!*

How many pals stay on surfboards?

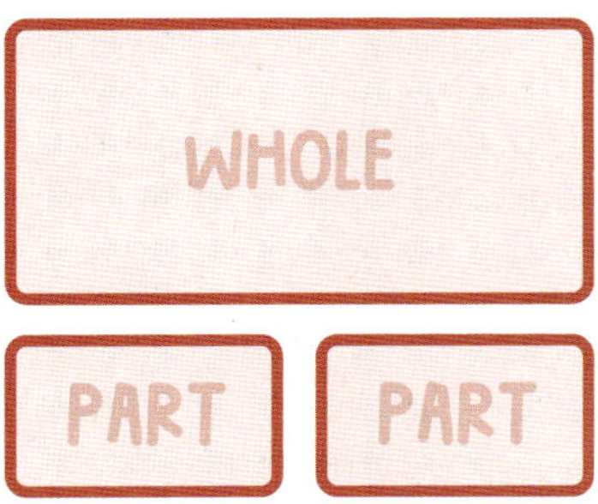

$4 - 2 =$ ______

Place the number stickers onto the correct spots in the part-part-whole model. Write the answer to the question.

The pals dry off.

Kazoo and Tali decide to make a sandcastle.

Kazoo fills 10 pails with sand.

Tali uses 4 pails of sand to make the castle.

How many pails of sand are left?

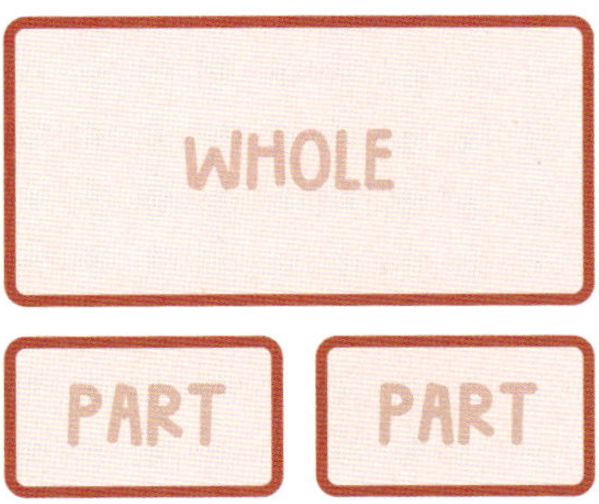

10 - 4 = ______

Place the number stickers onto the correct spots in the part-part-whole model. Write the answer to the question.

Now Chow, Zola, and Kazoo want to fly kites.

"My kite will fly the best," Kazoo says. "It has the longest tail."

His kite goes up. Its tail has 13 bows.

But part of the tail breaks off.

"Oh, no!" Kazoo yells. Now his kite tail only has 4 bows.

How many bows fell off?

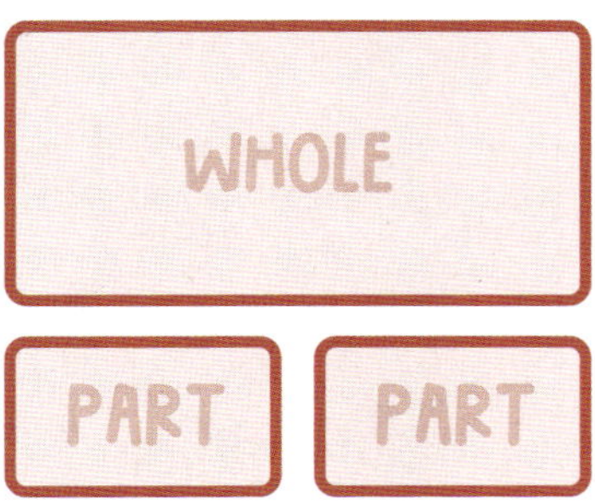

$13 - 4 =$ ______

Place the number stickers onto the correct spots in the part-part-whole model. Write the answer to the question.

The kite tail lands on the sand.

The crab picks it up and brings it to Kazoo.

Just then Batu returns. He holds trays with 8 ice cream cones.

“Would you like an ice cream?” he asks the crab. “It’s mint chip.”

Batu gives ice cream cones to all 5 friends.

How many are left for him to eat?

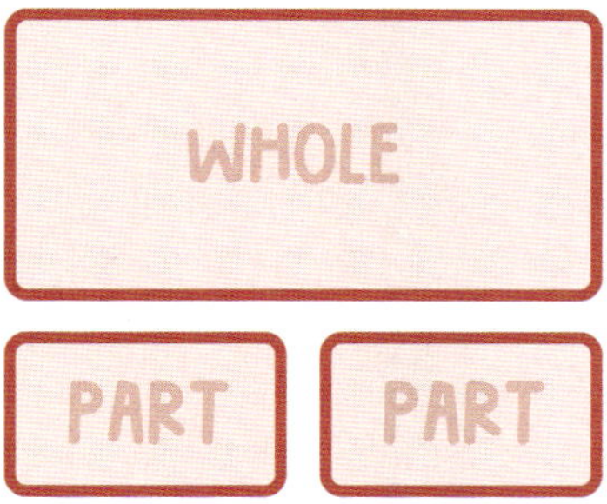

$8 - 5 =$ ______

Place the number stickers onto the correct spots in the part-part-whole model. Write the answer to the question.

Batu eats the rest of the ice cream cones.

"Mint chip is the best flavor ever," he declares.

"I think so, too," says the crab. "I'm glad we bumped into each other today."

HOOKED ON MATH

Steps 6 & 7

18023 (05/22

Math Fundamentals

The Alien Picnic

Written by Donna Giachetti
Illustrated by Aram Song

The Alien Picnic

Written by Donna Giachetti
Illustrated by Aram Song

Design by Bethany Schadel

Published by Sandviks HOP, Inc.

Printed in China.

18019

Here are 7 aliens, ready for some fun.
See how their spacesuits glow in the sun?

"It's a perfect day for a picnic," says Glip.
"But 1 of us needs to stay with the ship!"

It’s Glup’s turn today, they all decide.
Glup unhappily returns inside.

How many aliens go on a picnic?

7 - 1 = 6

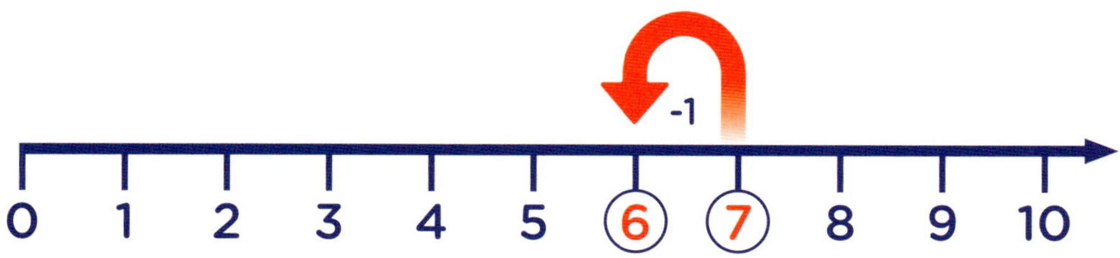

"I think this spot looks fine," says Gluf.
"Let's set our blanket on this green stuff."

Some aliens unpack tasty bites,
while 4 others head off to fly kites.

They've traveled far and want to play.
But—oh, no—3 kites fly away!

How many kites are left?

4 - 3 = 1

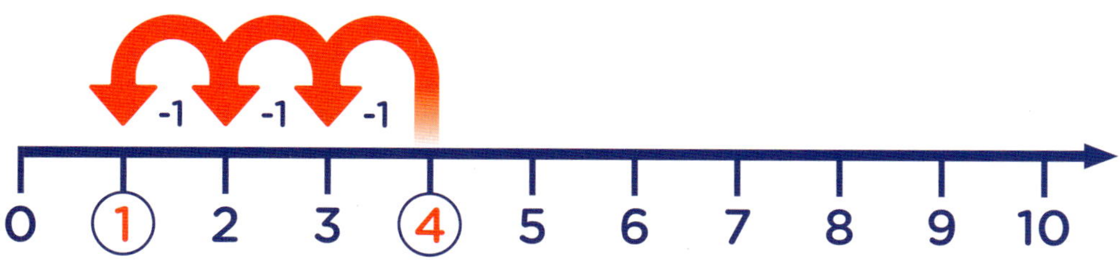

A platter of space eggs is passed to Glaf,
who chooses 1 and chomps it in half.

Now the plate has just 8 eggs.
“Save some for me!” Glek loudly begs.

Glek grabs 4 eggs and greedily munches.
She really loves these yummy lunches.

How many space eggs are left?

8 - 4 = 4

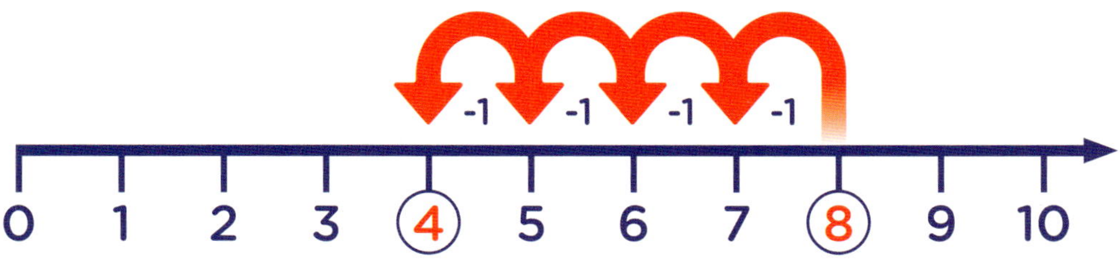

At first, the aliens don't even spy
the 9 little ants that happen by.

Until Glur stops playing his astro-fiddle
to loudly announce, "Look! Humans are little!"

Now the melon is moving!
What's happening here?
Then 7 tiny ants help the
snack disappear.

How many ants stay
on the blanket?

9 - 7 = 2

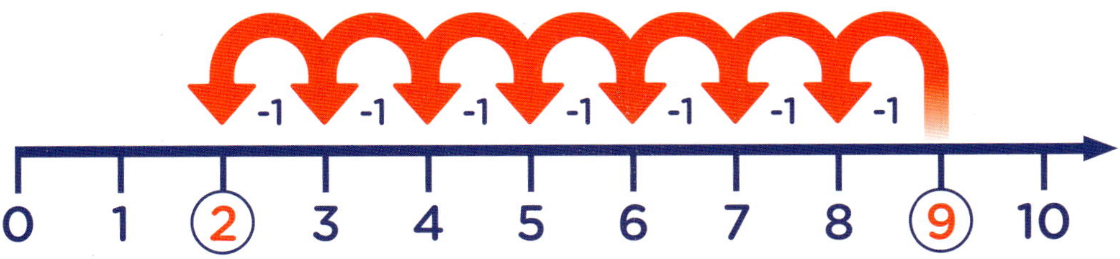

Glip lifts her face from a bowl filled with gloop,
an icy sweet type of space pumpkin soup.

"Join us, dear humans, there's plenty for all.
We don't mind, since you're really quite small."

So 10 ants climb up the stack so high.
But only 3 ants give the soup a try.

How many ants *don't* taste the gloop?

10 - 3 = 7

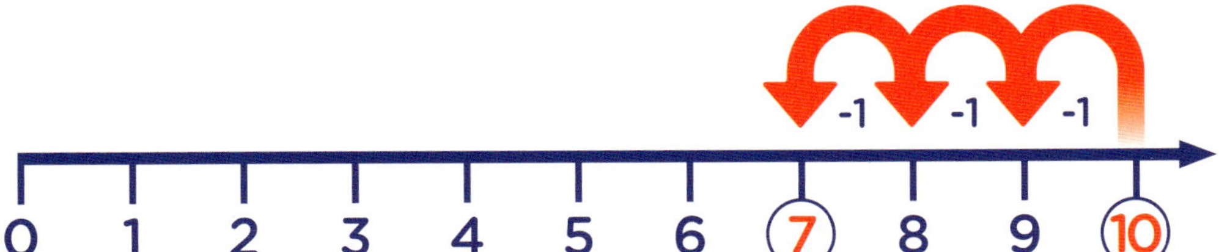

Yummy gloop, there's never enough!
Glek can't get her fill of the stuff.

Only 5 bowls left of the original 10.
Glek doesn't mind if she dips in again.

She gobbles 5 bowls, which seems quite a lot!
But Glek knows that gloop really hits the spot.

How many bowls of gloop are left?

$5 - 5 = 0$

The ants and aliens share the day,
making new friends the old-fashioned way.

Smiling Glof brings helmets into view.
"Would any of you like to join our crew?"

First, 7 ants quickly glance around,
then 5 ants decide to stay on the ground.

How many ants accept
the space helmets?

$7 - 5 = 2$

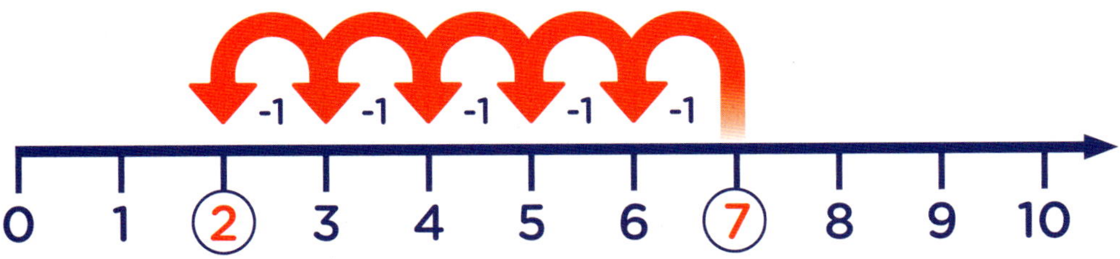

And so they gather, all ants in the group,
to say goodbye to some of their troop.

Then 2 ants put on helmets—just the right size.
Now they are ready to take to the skies.

The ship lifts off—up, up, and away!
Aliens and ants will be back one day.

HOOKED ON MATH

Steps 4 & 5

18019 (05/22)

Math Fundamentals

Chow the Magician

Written by Chris Sawyer
Illustrated by Nadeem Zaidi

Note to Parents

In this storybook, you will come upon many fact families like this one:

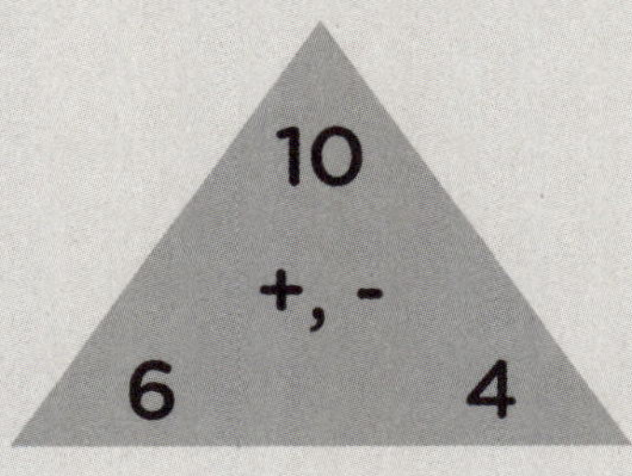

$6 + 4 = 10$
$4 + 6 = 10$
$10 - 6 = 4$
$10 - 4 = 6$

Facts about Fact Families

- Fact families are extremely useful for learning addition and subtraction.
- They are always made up of three related numbers.
- When the two smaller (bottom) numbers are added together, their sum equals the largest (top) number.
- When one of the smaller numbers is subtracted from the largest number, the difference is the other smaller number.

Chow the Magician

Written by Chris Sawyer
Illustrated by Nadeem Zaidi

Design by Bethany Schadel

Published by Sandviks HOP, Inc.

Printed in China.

18020

Chow saw a magic show today.
It was exciting!

It's time for bed now. But Chow is not sleepy.

He closes his eyes and counts sheep.

Chow imagines 7 sheep in a line.

They are waiting to see a magic show.

Oops! There is 1 sheep who forgot her ticket.

She goes back to get it.

How many sheep are left?

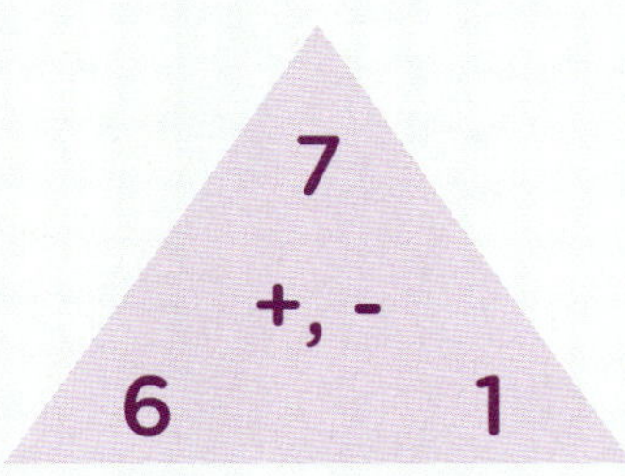

6 + 1 = 7

1 + 6 = 7

7 - 6 = 1

7 - 1 = 6

The Great Chowdini is starting his magic act.

He puts his top hat on the table.

He calls for his helper. “Hop to it!” he booms. “Zip! Zap! Zop!”

Poof! A bunny hops out of the hat! Then a second one. Then a third!

Oops! Now there are too many helpers.

Chow waves his wand. "Zap-a-roo, 2 of you!"

How many bunnies are left?

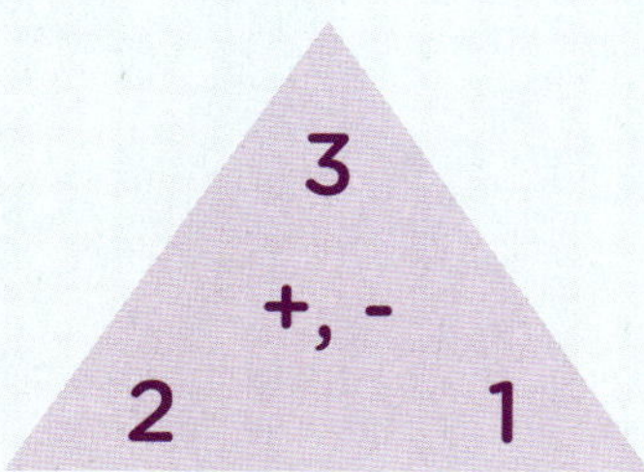

2 + 1 = 3

1 + 2 = 3

3 - 2 = 1

3 - 1 = 2

The helper tosses the Great Chowdini 6 balls.

"I will now make these balls disappear!" he says.

He tries to juggle. Oops! He drops 2 of the balls.

Boing! They bounce off the stage.

"Uh, ta-da!" the Great Chowdini says.
"I made some balls disappear!"

How many balls are left in the air?

6
+, -
4 2

4 + 2 = 6
2 + 4 = 6
6 - 4 = 2
6 - 2 = 4

Chow invites Zola onto the stage. He tells her to draw some dots.

“The Great Chowdini will guess the number of dots,” he says.

Chow thinks. “You drew 9. I’m sure of it!”

Zola shakes her head. “I drew 5 dots fewer than that,” she tells him.

How many dots did Zola draw?

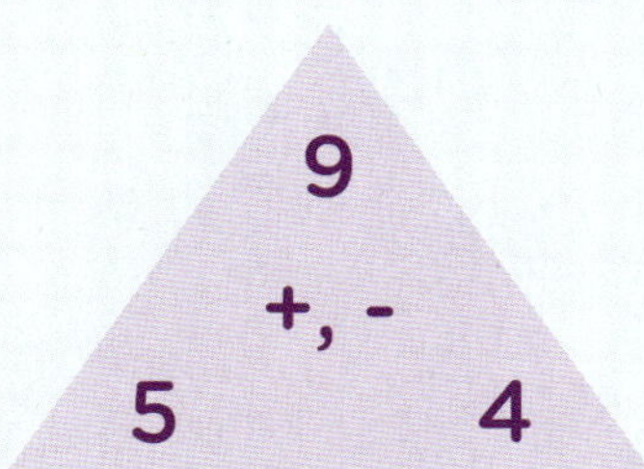

5 + 4 = 9
4 + 5 = 9
9 - 5 = 4
9 - 4 = 5

The Great Chowdini pulls 8 scarves out of a magic box.

“I will turn these scarves into flowers for you,” he tells Zola.

“How nice,” she says.

The Great Chowdini waves his wand.
"Bop! Hop! Zop!"

None of the scarves turn into flowers.

But 3 of them turn into frogs!

The frogs hop all over the stage.

A frog even hops onto Zola's head!

How many scarves are left?

8

+, -

5 3

5 + 3 = 8
3 + 5 = 8
8 - 5 = 3
8 - 3 = 5

"For my next trick, I will make Zola vanish," the Great Chowdini announces.

"You will?" Zola asks him.

Before Zola can disagree, he drapes a curtain over her.

He shouts, "Sha-zing!"

Poof!

When the smoke clears, Zola is still there.

But where are her spots?

"I had 4 spots," she cries. "Now I only have 1!"

Oops!

How many spots did Zola lose?

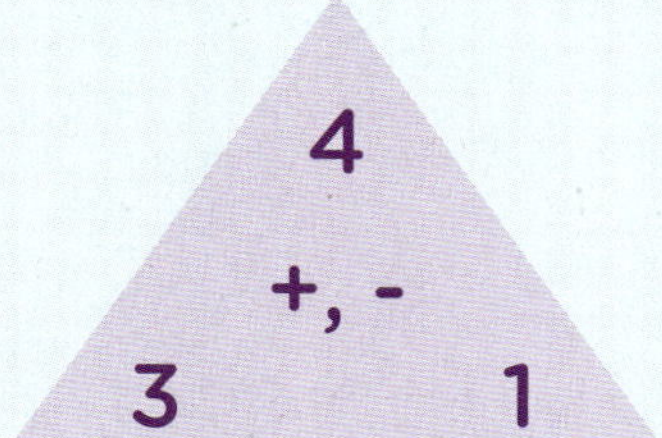

3 + 1 = 4

1 + 3 = 4

4 - 3 = 1

4 - 1 = 3

"For my final trick, I will make 10 balloons dance," he booms.

The Great Chowdini takes a deep breath.

He blows into his magic trumpet. *Tooooooot!*

The noise is so loud, the balloons pop!

Oops! Wait, what's that?

Zola floats up from the box!

She is tied to the 1 balloon that did not pop.

But she is happy. She has her spots back!

How many balloons popped?

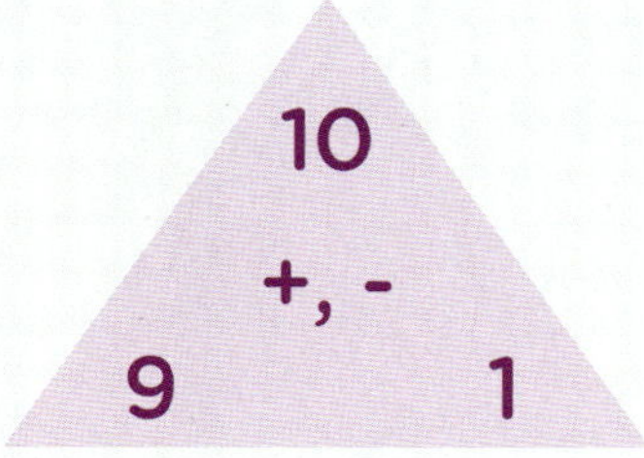

9 + 1 = 10
1 + 9 = 10
10 - 9 = 1
10 - 1 = 9

Chow wakes up to a bright new day.

"I was not a very good magician. I'm glad that it was all just a dream," he says.

Or was it?

HOOKED ON MATH

Steps 4 & 5

18020 (05/22)

HOOKED ON MATH

Math Fundamentals

Steps 1, 2, & 3

Workbook

Tips for Success

- Math Fundamentals introduces your child to the building blocks of math—number sense, addition, and subtraction.
- Work at a pace that's right for you and your child.
- Fit your session length to your child's attention span. It's okay to work with the App one day and this workbook the next.
- Before beginning, flip through the workbook and note any materials you might need. You should plan to have pencils, a pair of dice, and crayons ready.
- You'll see parent tips throughout this workbook. These tips are designed to answer questions you or your child might have about the learning.
- Repetition and review are essential to basic math fluency. Each workbook step builds on the ones before, so make sure your child has mastered the current lesson before moving forward.
- As you go through the workbook, encourage your child to explore the connection between numbers and the real world. Reinforce concepts with familiar household objects such as toys or snacks to make adding and subtracting more memorable.
- Have fun with it!

HOOKED ON MATH

Math Fundamentals

Steps 1, 2, & 3

Workbook

This workbook and the included storybooks introduce fundamental math skills.

Step 1: Counting Numbers to 20
- Counting 1 to 20
- Ordering numbers

Step 2: Adding to 10
- Adding to 10
- Number sentences
- Adding with 0

Step 3: Addition Practice
- Addition facts to 10
- Vertical addition

Contents

Lesson Overview

Welcome! This workbook provides supplemental activities that will solidify and expand the learning your child is enjoying with the Hooked on Math App. Follow the easy, step-by-step instructions to guide your child's exploration of numbers and operations. Every lesson helps your child build a strong foundation in fundamental math skills while having fun!

1. Launch

Your child will be introduced to math concepts and skills through interactive lessons in the App on your mobile device or Windows or Mac computer. Download the App at:

iOS

hookedonmath.com/get-the-app

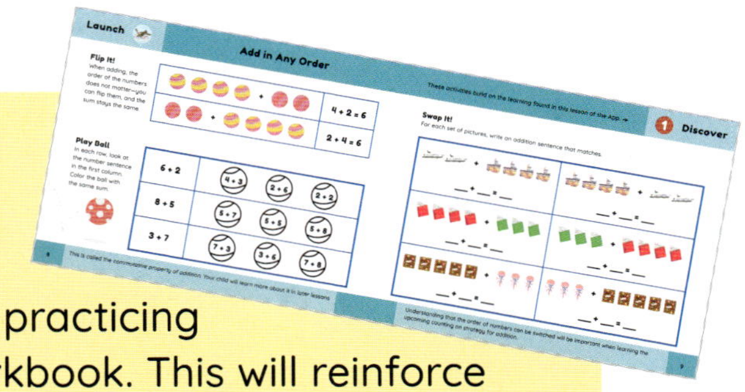

2. Discover

Next, your child will have fun exploring math concepts and practicing new skills in the workbook. This will reinforce the App lessons and provide a tangible application of the skills learned.

3. Play

In addition to the workbook activities, your child will enjoy two math-based storybooks that will strengthen their learning and add to the fun!

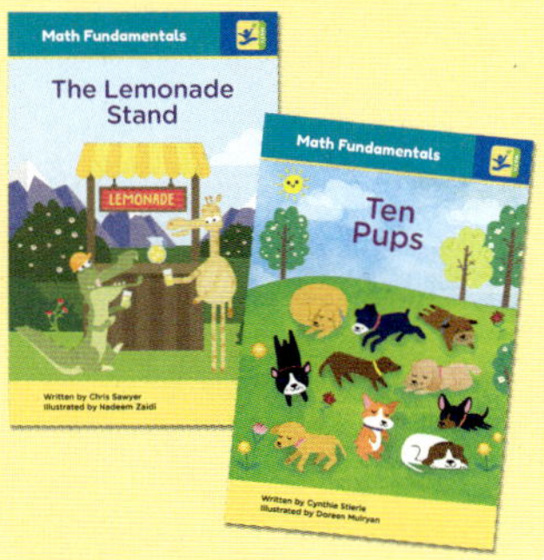

4. Celebrate

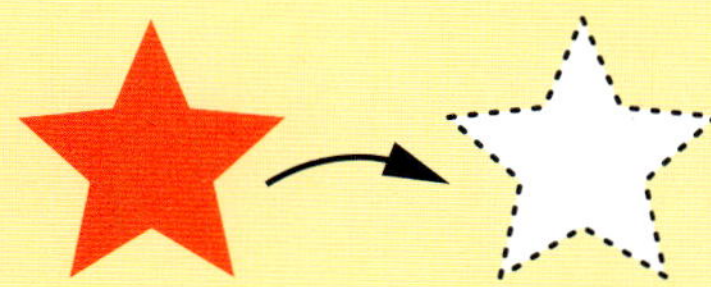

Celebrate your child's progress by placing a sticker in the workbook as learning milestones are reached. This also provides a helpful marker to remember where to begin next time.

Special notes are included in this space to provide additional context related to what is being taught.

Counting Forward and Backward to 10

Counting is naming numbers in order. When counting forward, the numbers go up one by one.

Forward March

Look at the chart below.

- Say the number names.
- Color and count the stars.
- Read the number words.
- Trace each numeral.

1	2	3	4	5	6	7	8	9	10
☆	☆☆	☆☆☆	☆☆☆☆	☆☆☆☆☆	☆☆☆☆☆☆	☆☆☆☆☆☆☆	☆☆☆☆☆☆☆☆	☆☆☆☆☆☆☆☆☆	☆☆☆☆☆☆☆☆☆☆
one	two	three	four	five	six	seven	eight	nine	ten
1	2	3	4	5	6	7	8	9	10

It's important for your child to understand the idea that a number represents a specific quantity of objects. Encourage your child to count each star as it is colored.

When counting backward, the numbers are named in reverse order and go down one by one.

Blast Off!

Help the astronaut count down to blast off.

- Fill in the missing numbers.
- Draw the correct number of stars.
- Write the missing number words.
- Trace each numeral.

10		8		6	5		3		1
☆☆☆☆☆☆☆☆☆☆	☆☆☆☆☆☆☆☆☆		☆☆☆☆☆☆☆	☆☆☆☆☆☆		☆☆☆☆		☆☆	☆
	nine	eight	seven		five	four	three	two	
10	9	8	7	6	5	4	3	2	1

A *digit* is any numeral from 0 to 9. Point out to your child that the numeral 10 is not the same as the others because it is a two-digit numeral.

Launch

Counting Numbers 11 to 15

After the number 10, the numbers begin to follow a pattern.

The number **11** is 10 plus 1 more.

11 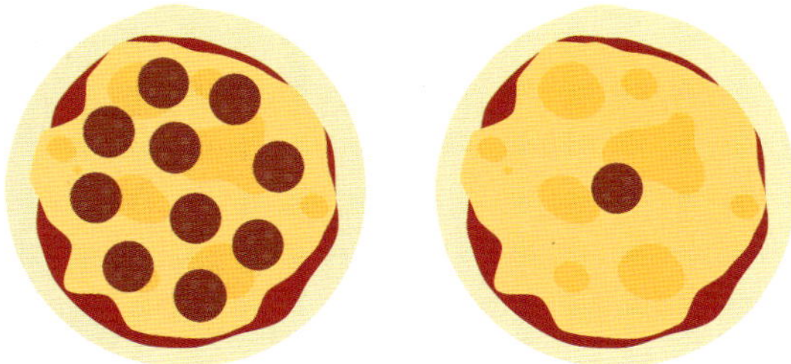

The number **12** is 10 plus 2 more.

12 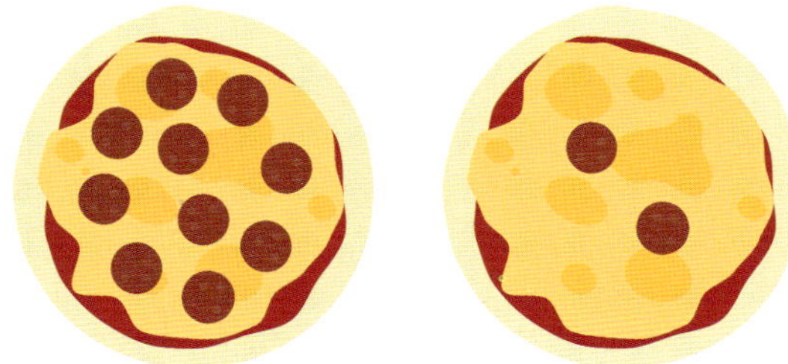

The number **13** is 10 plus 3 more.

13 

The number **14** is 10 plus 4 more.

14 

The number **15** is 10 plus 5 more.

15 

Shell Game

Count the shells in each group, then draw a line to the matching number.

Tip: Use a pencil to make an **X** through each shell as you count it.

11

12

13

14

15

Numbers

Numbers are used to calculate, count, or measure. They are abstract and can be expressed in many ways: as words (five); sets (• • • • •); or symbols (like the numeral 5).

Counting Numbers 16 to 20

The pattern continues with the rest of the teen numbers.

The number **16** is 10 plus 6 more.

16 

The number **17** is 10 plus 7 more.

17 

The number **18** is 10 plus 8 more.

18 

The number **19** is 10 plus 9 more.

19 

The number **20** is 10 plus 10 more.

20 

Make sure your child has a solid understanding of numbers and what they represent. Use daily experiences to relate quantities of objects to numbers. Counting snacks, toys, or traffic lights are all opportunities to develop number sense.

Treasure Hunt Maze

Help Zola find the treasure chest by coloring the numbers in the bubble maze in order from 1 to 20.

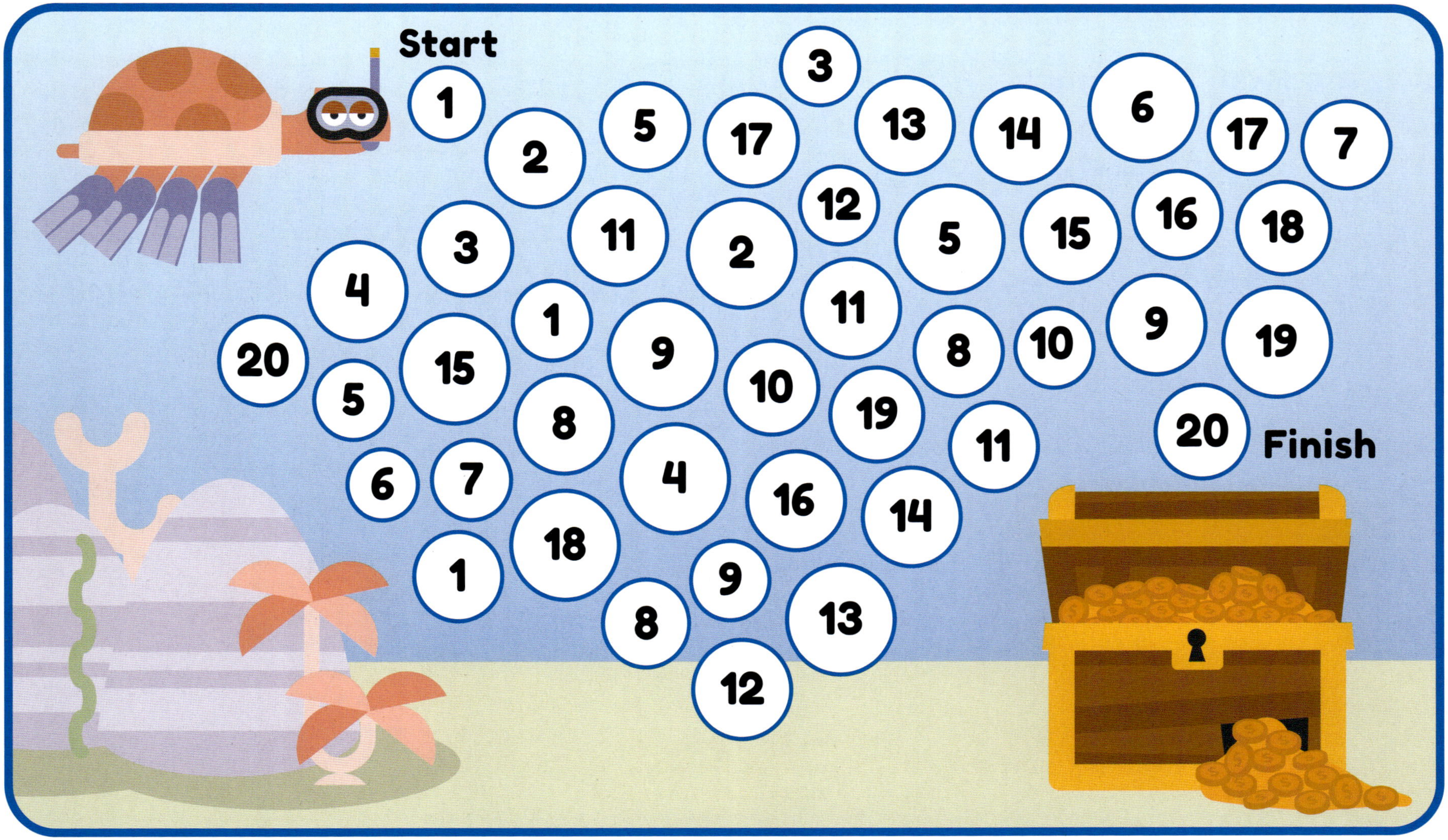

Launch

Before and After

When you order numbers, you place them in a list.
There are numbers that come *before* other numbers. For example, 3 comes before 4.
There are numbers that come *after* other numbers. For example, 4 comes after 3.

Using a Number Line

To help with ordering numbers, you can use a tool called a *number line*. It is a straight line with numbers placed at equal spaces, or segments, along its length.

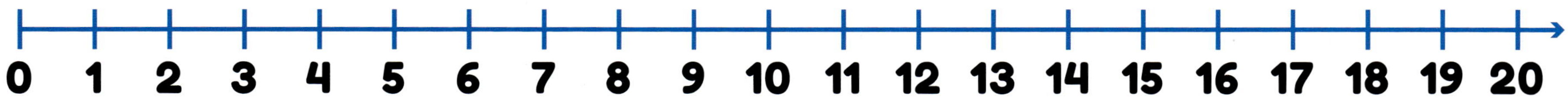

Chow Time

Fill in the missing numbers along Chow's back.
Use the number line to help you.

Caterpillar Crawl

Fill in the missing numbers on each caterpillar.
Use the number line to help you.

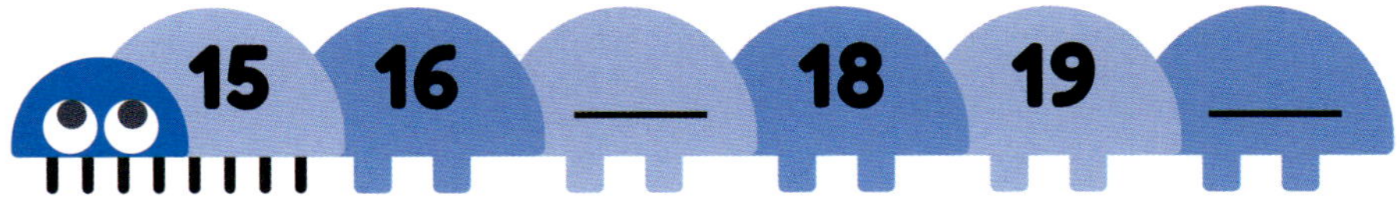

Before	
11	12
	15
	3
	9
	20

After	
16	17
6	
11	
5	
2	

Line Up!

Use the number line to help you find the number that is right before or after the number shown in each column. The first one has been done for you.

A child who is able to put numbers in order has grasped the *stable-order principle*, which means that numbers are in a fixed sequence that never changes.

Comparing Numbers

Math uses numbers and symbols called *signs* in place of words to write sentences that are read from left to right. Some signs compare numbers.

This sentence tells you that 9 is more, or *greater than*, 5.	**9 > 5**
This sentence tells you that 12 is fewer, or *less than*, 20.	**12 < 20**
This sentence tells you that both amounts are the same, or *equal to* each other.	**10 = 10**

>	greater than
<	less than
=	equal to, or the same

See the Signs

Compare the groups of objects and write the correct symbol in the circle to make the math sentence true.

Compare-a-Spin

Spin the first spinner and write the number in the first column. Spin the second spinner and write the number in the second column. Then fill in the center circle with the symbol that makes the sentence true.

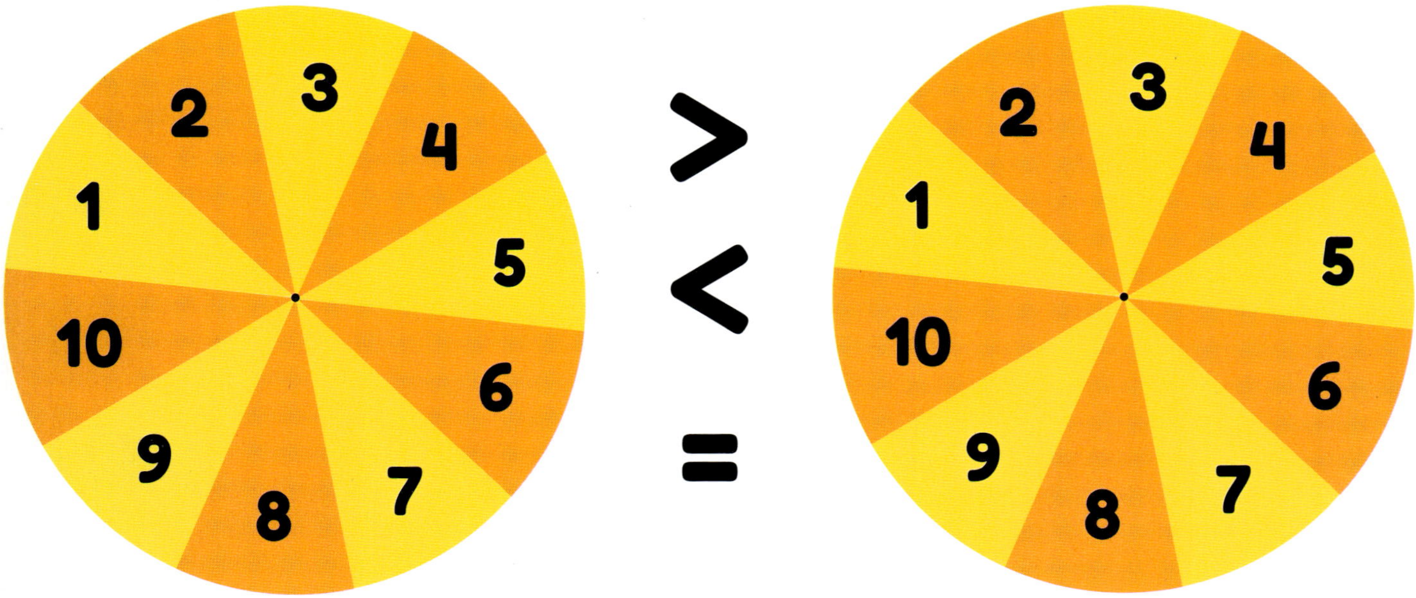

3	<	10

How to Make the Spinner: Place one end of the paper clip in the center of the circle. Put the pencil tip inside the paper clip on the center dot. Now flick the paper clip so it spins around the pencil.

Materials needed: sharpened pencil, paper clip

FIN-ish the Friend

Who is swimming in the water? Follow the numbers in order from 1 to 20 to connect the dots and find out. Then color the picture.

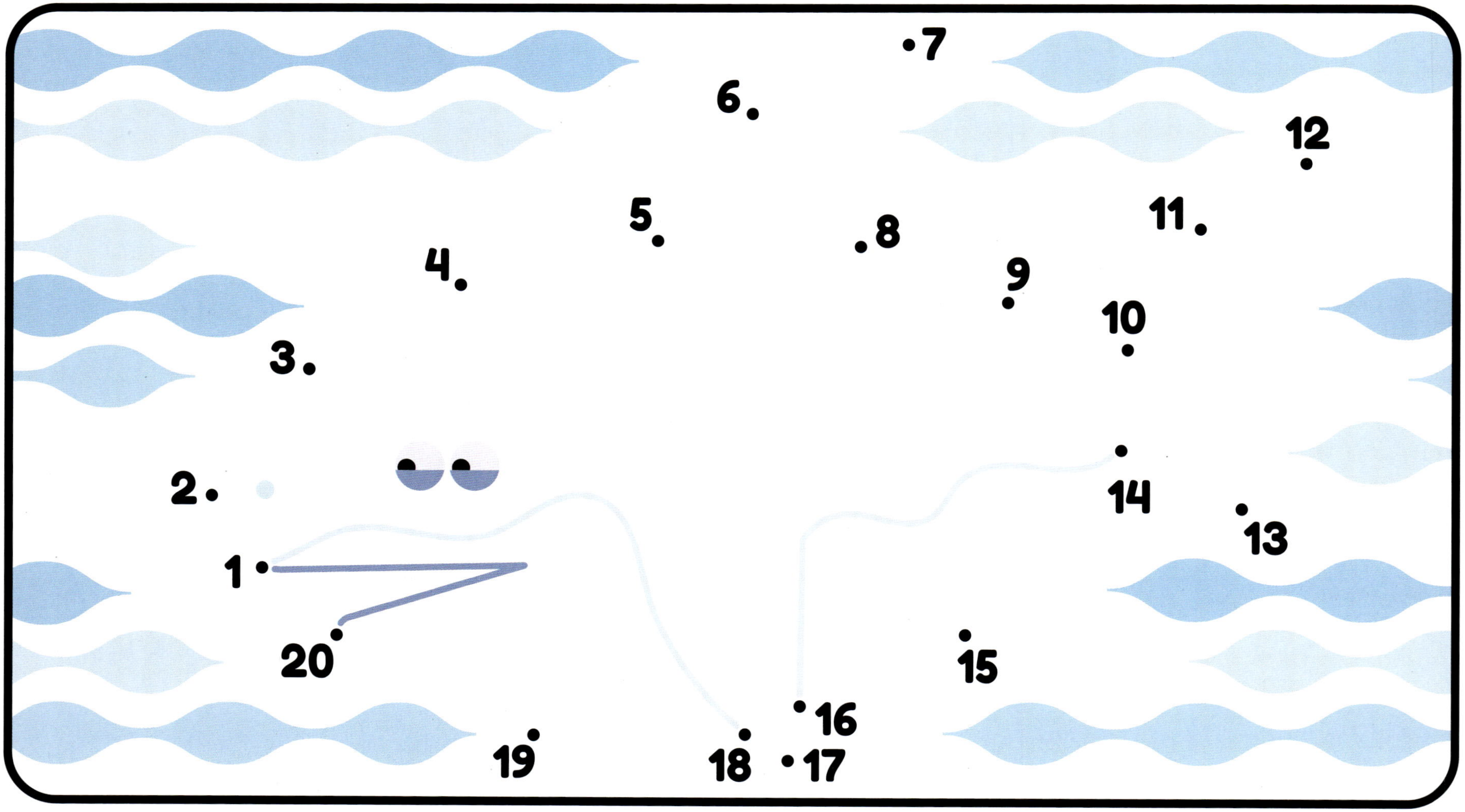

Swimming Schools

- Count the fish in each group. Write the number in the box.
- Circle the group with the *greatest* number of fish.
- Draw a box around the group with the *least* number of fish.
- Make a check mark (✓) next to the two groups that have an equal number of fish.

Launch

Introducing Addition

Addition is putting two or more groups together to find how many in all.

Read and Count

Read the stories. Count the items to add and then write the answer in the box.

The bird lays 1 egg in her nest. Then she lays 2 more. How many eggs are in the nest in all?

The clown is juggling 3 balls. Then he adds 2 more balls. How many balls is the clown juggling in all?

Tali bakes 6 muffins. Then she bakes 6 more. How many muffins does she bake in all?

Kip spies 6 ladybugs on a twig. Then he sees 4 more. How many ladybugs does he see?

The *cardinal principle* is the understanding that when counting objects, the last number counted is the total number of objects in the group.

Draw the Story

Read the addition stories below. Then draw all the items from the story in the empty space. Count all the items and write the answer to the question.

Batu sees 4 stars in the sky. Then he spots 2 more. How many stars does he see?

☐

Zola reads 3 books. Then she reads 3 more. How many books does she read?

☐

Kazoo picks 5 apples. Then he picks 3 more. How many apples does he pick?

☐

Understanding Addition

The symbol for addition is called the *plus sign* (+). It tells you to combine the groups to find out how many there are in all.

Add It

Count the objects to add the groups together. Circle the number that tells how many there are in all.

6 8 9

11 15 20

7 8 10

Keep Rolling

- Roll two dice. Draw the dots for each one.
- Count the dots and write the number for each one.
- Count both sets of dots and write how many in all.

The first one has been done for you.

Roll	Draw Die 1	Count	Draw Die 2	Count	How many in all?
Roll 1	● ● ● ● ● ●	6	● ● ● ● ●	5	11
Roll 2					
Roll 3					
Roll 4					
Roll 5					
Roll 6					

Materials needed: Pair of dice

Launch

Addition Number Sentences

Math Talk

An *addition sentence* is made up of two or more numbers added together. Their total is called the *sum*.

The signs in a math sentence tell you what mathematical action, or *operation*, you need to do.

- The *plus sign* tells you to add the numbers together.
- The *equal sign* tells you that the quantity on one side of the sign is the same as the other side.

Plus sign → +; Sum → 7; Equal sign → =

$$4 + 3 = 7$$

See the Signs

Look at the math sentences below. Underline the numbers to add. Circle the sums. Color the plus signs **blue**. Color the equal signs **red**. The first one has been done for you.

$1 + 4 = 5$ $\quad$ $7 + 2 = 9$ $\quad$ $5 + 5 = 10$

$6 + 1 = 7$ $\quad$ $3 + 5 = 8$ $\quad$ $1 + 8 = 9$

The two numbers that are added together are called *addends*.

It Adds Up

Draw a line to match the addition picture with its addition sentence. Then match the picture to the correct sum. The first one has been done for you.

Add Together	Picture	Sum
3 + 2		9
2 + 2		8
7 + 1		5
6 + 4		10
5 + 4		4

Adding with 0

The number 0 stands for nothing. When adding 0 to any number, it does not change the number.

zero

zero

$$1 + 0 = 1$$

Nothing to Read

Trace the word above. What does it say? Write that number inside the starfish.

Nothing There

Look at the picture pairs below. Circle the picture in each pair that shows the number 0, or nothing.

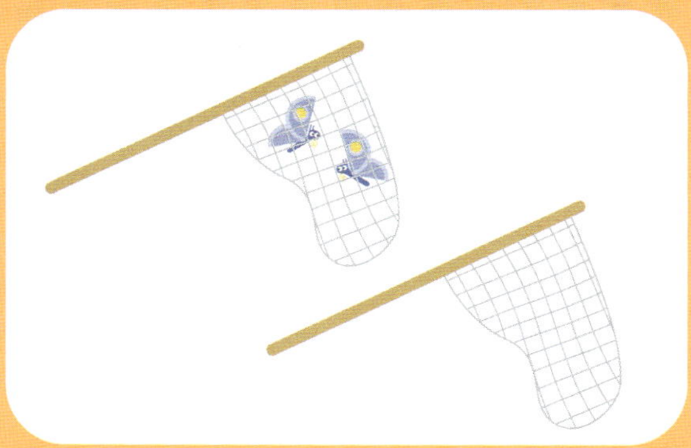

Explain to your child that while zero represents nothing, it is a very important number in math. It is a placeholder. Without it, we would not be able to represent numbers like 10, 100, or 1,000.

Ring Around Zero

Add the 0 in the center ring to each of the numbers around it. Write the answer in the outer ring. The first one has been done for you.

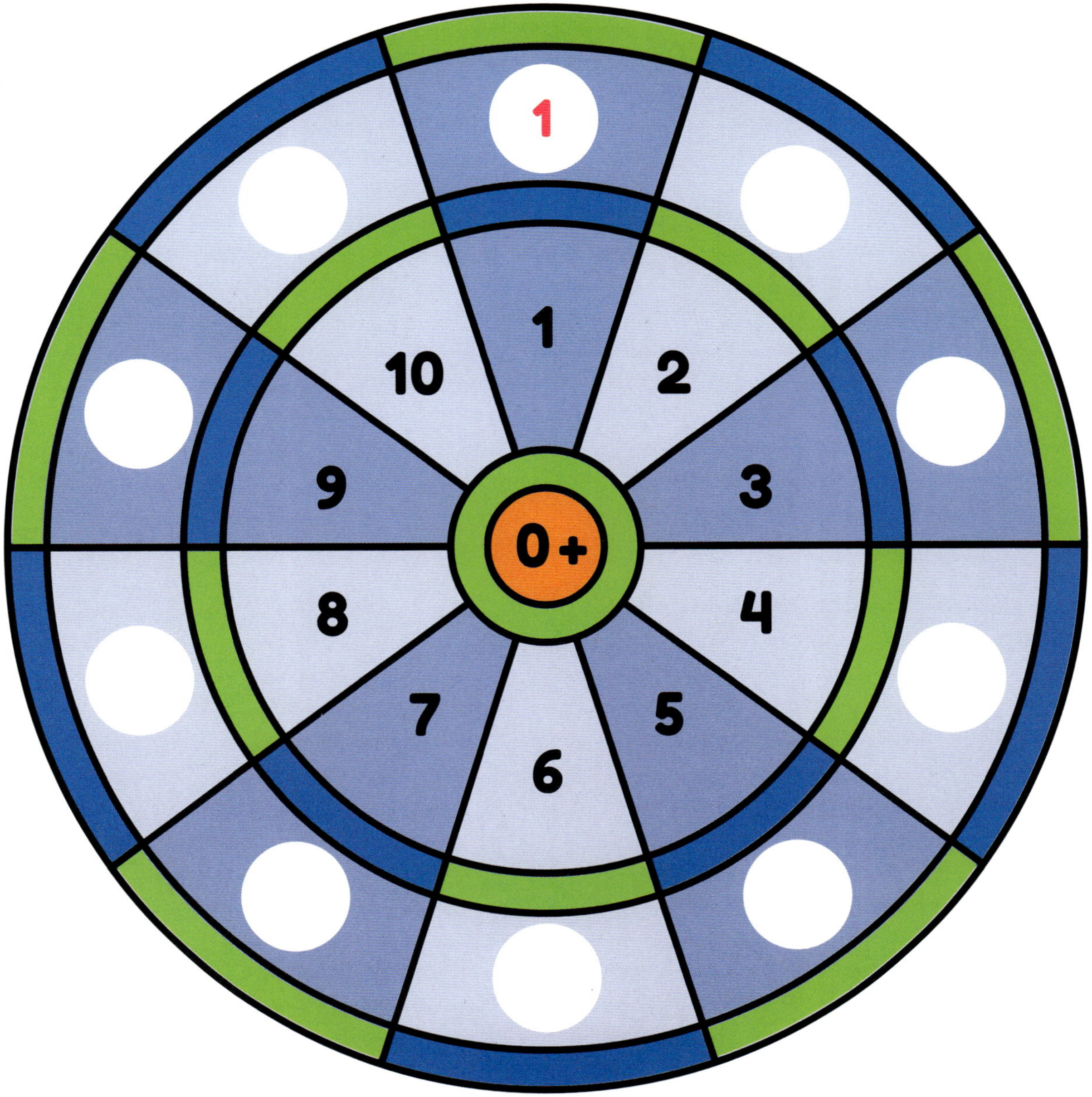

Riddle: What did 0 say to 8?

Answer: I like your belt!

Fact Families

A *fact family* is made up of three numbers that are related to each other.

When the two smaller numbers are added together, they equal the largest number.	6 + 4 = 10 4 + 6 = 10
When one of the smaller numbers is taken away from the largest number, it equals the other smaller number.	10 - 4 = 6 10 - 6 = 4

10

+, -

4 6

Fact Family Find

Look at each fact family.

- Circle the two smaller numbers.
- Draw a box around the largest number.
- Draw a check (✓) next to each addition sentence.
- Draw an ✗ next to each subtraction sentence.

9

+, -

2 7

2 + 7 = 9
7 + 2 = 9
9 - 2 = 7
9 - 7 = 2

8

+, -

3 5

3 + 5 = 8
5 + 3 = 8
8 - 3 = 5
8 - 5 = 3

Subtraction is covered in future lessons, but it's important for your child to see that addition and subtraction are inverse operations. Mastering addition fact families will help with subtraction later.

Who's Home?

Each house shows one part of the fact family. Fill in the missing facts.

6

+, −

2 4

2 + ___ = 6
4 + ___ = 6
6 − 2 = 4
6 − 4 = 2

5

+, −

2 3

___ + 3 = 5
3 + ___ = 5
5 − 2 = 3
5 − 3 = 2

9

+, −

4 5

___ + 5 = 9
5 + ___ = 9
9 − 4 = 5
9 − 5 = 4

10

+, −

3 7

3 + ___ = 10
7 + ___ = 10
10 − 3 = 7
10 − 7 = 3

Code Breaker

Using the key, write each number in the box below its matching picture to write a number sentence.

Key

0	1	2	3	4	5	6	7	8	9	10

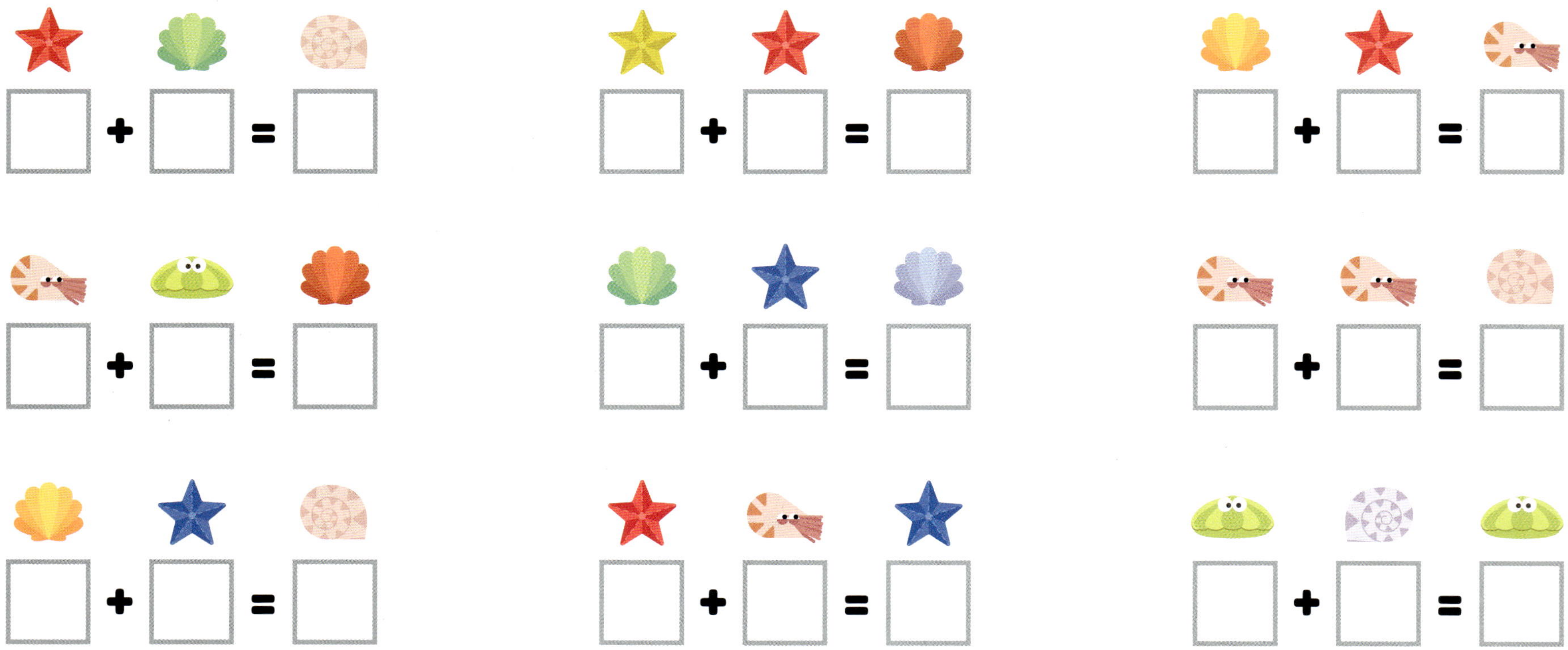

Color Key

If the sum is 1 or 2 = blue
If the sum is 3 or 4 = purple
If the sum is 5 or 6 = green
If the sum is 7 or 8 = yellow
If the sum is 9 or 10 = orange

Color by Number Addition

Find the sum of the two numbers in each spot. Then use the Color Key to finish the picture.

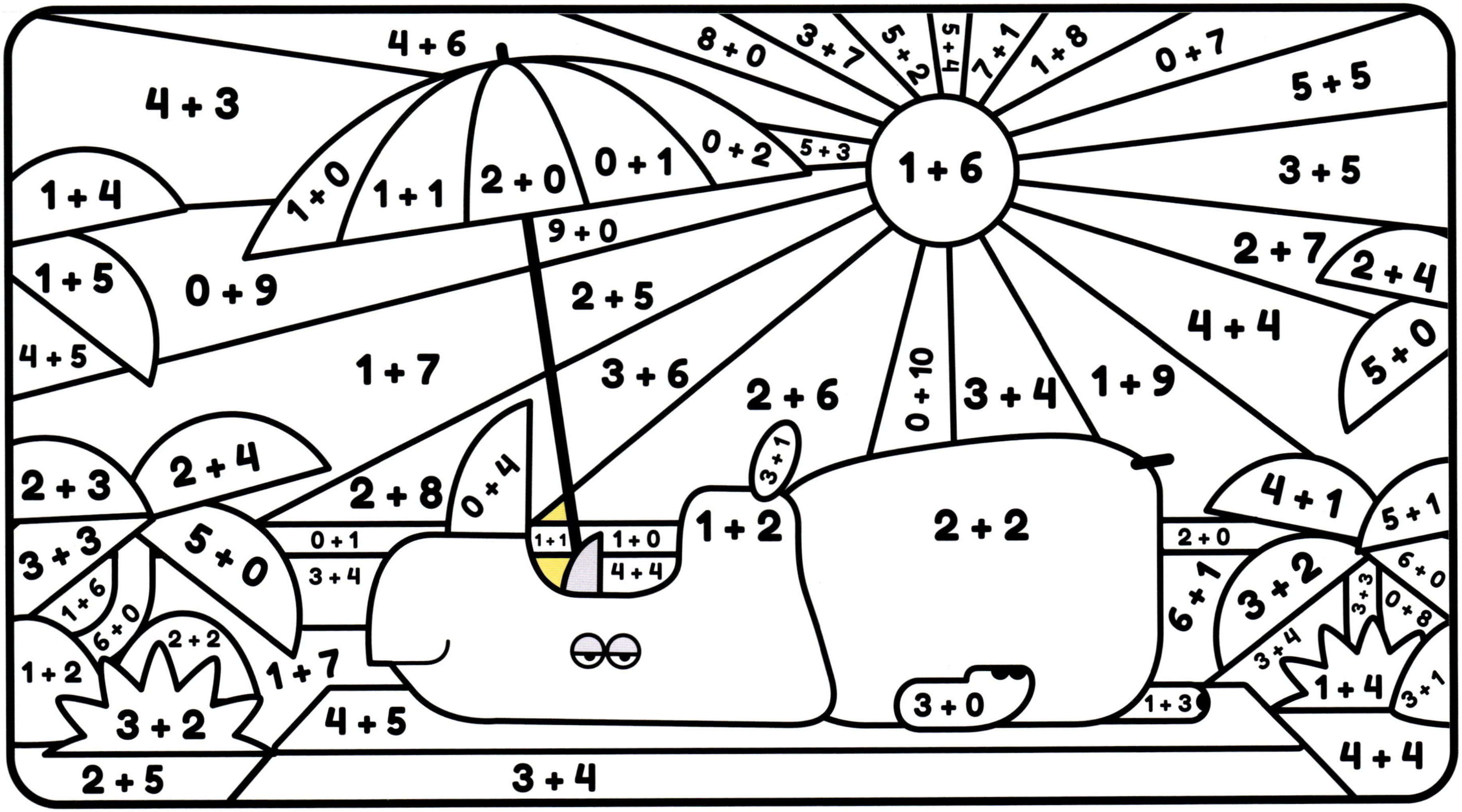

Read The Lemonade Stand

Read the storybook *The Lemonade Stand* with your child. This book includes hands-on activities to encourage your child to practice their addition skills. Use the sticker sheet marked "The Lemonade Stand" that came with this pack.

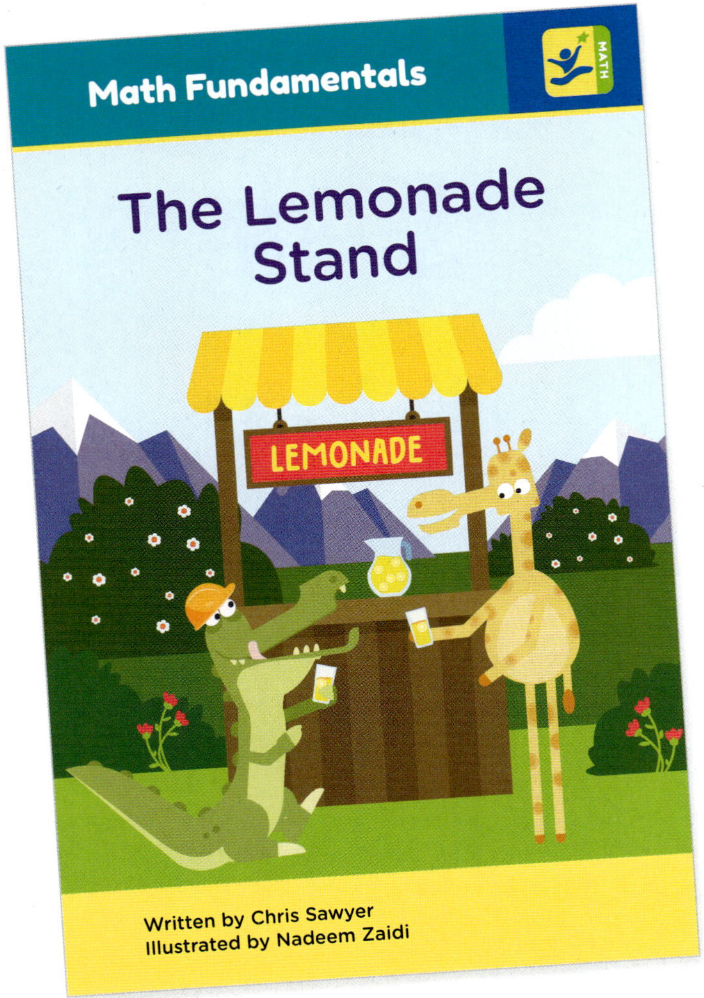

After reading, use these ideas to continue the learning:

- Make a pitcher of lemonade together! How many lemons will you use? How many cups of water? Count the ice cubes, too.
- Pretend to open a lemonade stand at home. How much will your child charge per cup? This activity is a great way to bring counting into a real-life situation.
- What foods does your child like to eat? Look at what you need to make a favorite recipe. Let your child measure and count out the ingredients the next time you are in the kitchen together.

Celebrate your child's progress by placing a star sticker in the workbook.

Move with Math

Celebrate by moving with math! In a safe, open area, try each action. Then write the answer to the number sentence.

Clap 3 times! Now clap 6 more times! How many times did you clap?	**3 + 6 = ___**
Jump 4 times. Now jump 5 more times! How many times did you jump?	**4 + 5 = ___**
Touch your toes 8 times. Now touch your toes 2 more times. How many times did you touch your toes?	**8 + 2 = ___**
Spin 3 times in a circle. Now spin 4 more times. How many times did you spin?	**3 + 4 = ___**

Your child has completed Step 2!
Celebrate by placing a star sticker in the workbook.

Ways to Make 4 and 5

FOUR 4

You can make the sum of 4 in different ways, using different addition sentences.

0 + 4 = 4	4 + 0 = 4
1 + 3 = 4	3 + 1 = 4
2 + 2 = 4	

Color the rainbow to connect the numbers that equal 4.

Color each part of the rainbow the same color as the two numbers that equal 4.

In an addition sentence, the numbers being added can be flipped. Both 1 + 3 and 3 + 1 equal a sum of 4!

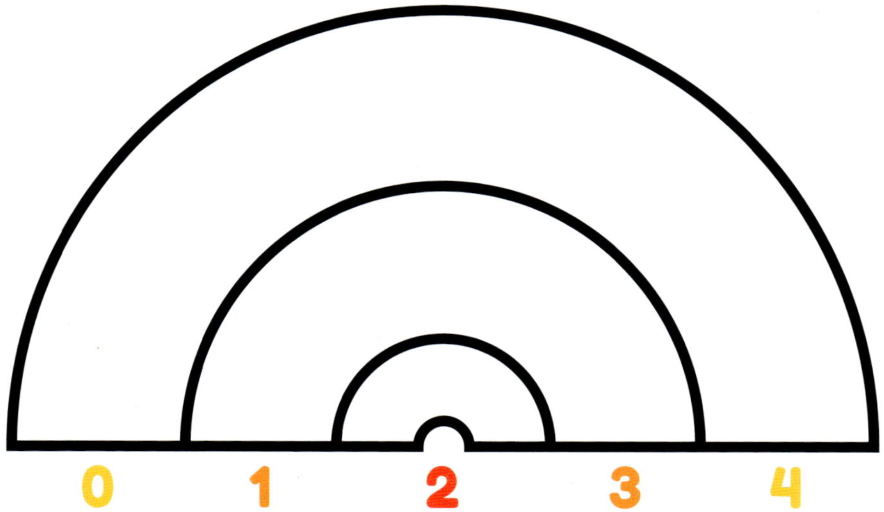

Notice that the number 2 is added to itself to equal the number 4. 2 + 2 = 4.

FIVE 5

You can make the sum of 5 in different ways, using different addition sentences.

0 + 5 = 5	5 + 0 = 5
1 + 4 = 5	4 + 1 = 5
2 + 3 = 5	3 + 2 = 5

Color the rainbow to connect the numbers that equal 5.

Color each part of the rainbow the same color as the two numbers that equal 5.

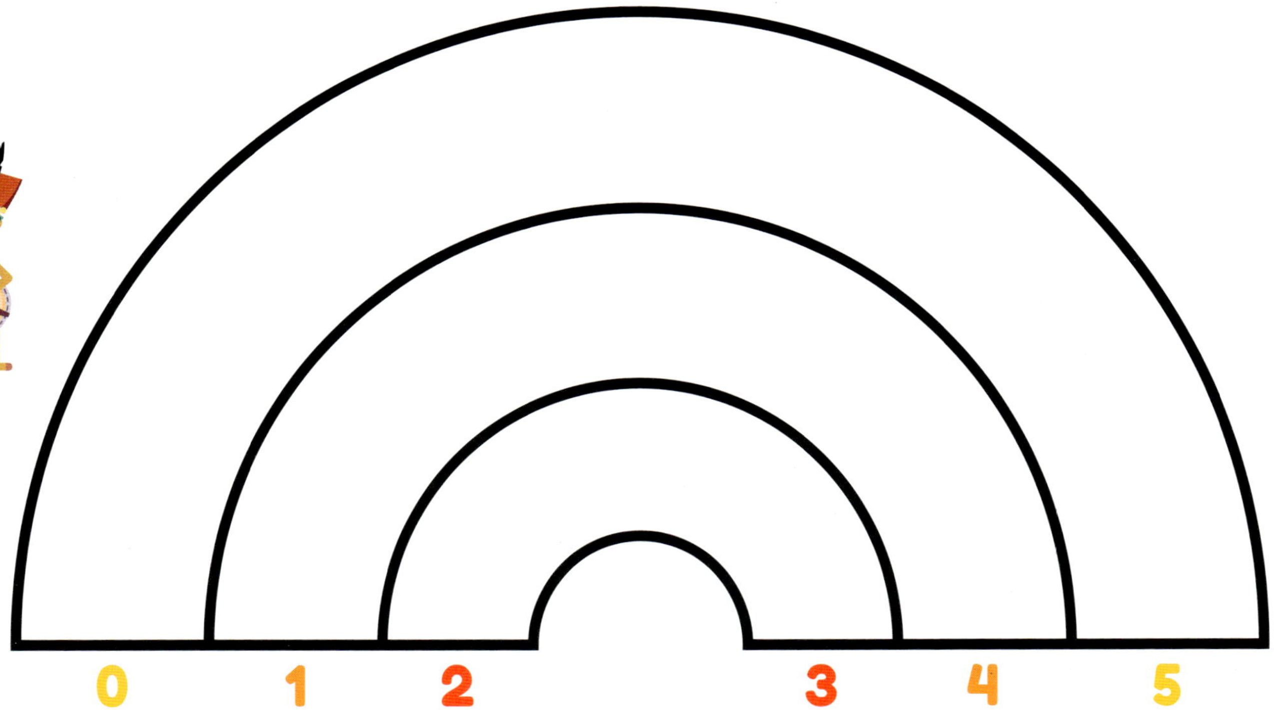

Materials needed: crayons or colored pencils in the colors of the rainbow for this and the upcoming pages

Sum It Up

Look at the addition sentences in the picture. If the two numbers equal **4**, color that part of the picture **green**. If they equal **5**, color that part of the picture **blue**.

Tic-Tac-Toe

Find the sums for each addition sentence.

- Draw an **O** around the sentences that equal 4.
- Draw an **X** through the sentences that equal 5.
- See which number wins in each game.

Riddle: Did 4 win the only prize?

Answer: No, 5-1-1-2.

0 + 5	1 + 3	4 + 1
3 + 2	1 + 4	0 + 4
2 + 2	4 + 0	2 + 3

5 + 0	0 + 5	3 + 1
1 + 4	1 + 3	2 + 2
4 + 0	3 + 2	2 + 3

Ways to Make 6 and 7

SIX 6

You can make the sum of 6 in different ways, using different addition sentences.

0 + 6 = 6	6 + 0 = 6
1 + 5 = 6	5 + 1 = 6
2 + 4 = 6	4 + 2 = 6
3 + 3 = 6	

Color the rainbow to connect the numbers that equal 6.

Color each part of the rainbow the same color as the two numbers that equal 6.

Math can be very creative! Gather a small group of objects, such as buttons or toy cars. Select a number and ask your child to arrange that number of objects in as many groupings as possible.

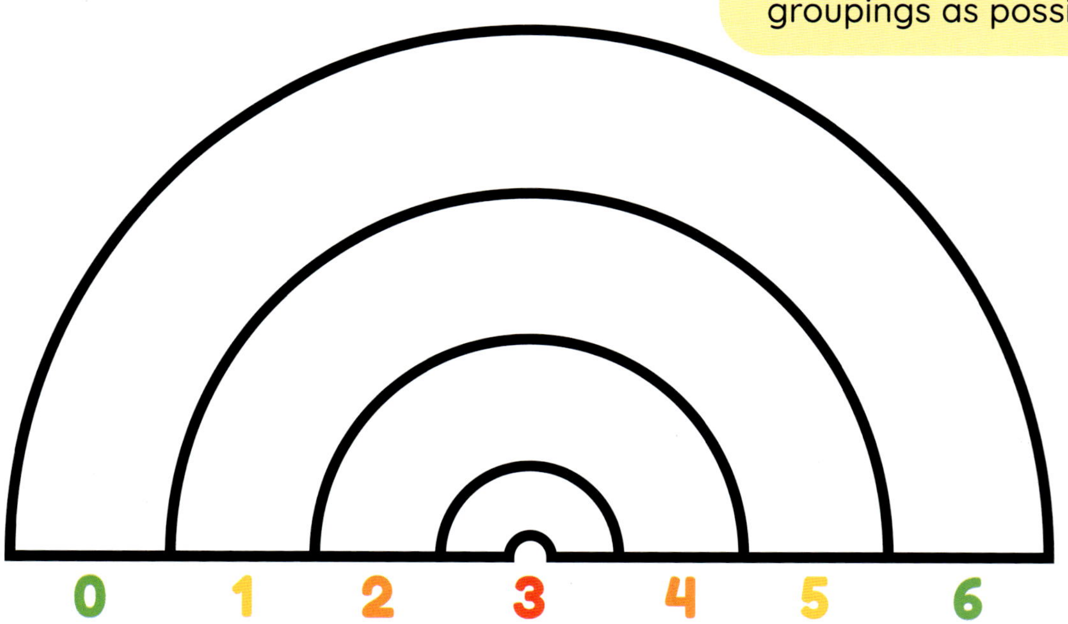

Notice that the number 3 is added to itself to equal the number 6.
3 + 3 = 6.

SEVEN 7

You can make the sum of 7 in different ways, using different addition sentences.

0 + 7 = 7	7 + 0 = 7
1 + 6 = 7	6 + 1 = 7
2 + 5 = 7	5 + 2 = 7
3 + 4 = 7	4 + 3 = 7

Color the rainbow to connect the numbers that equal 7.

Color each part of the rainbow the same color as the two numbers that equal 7.

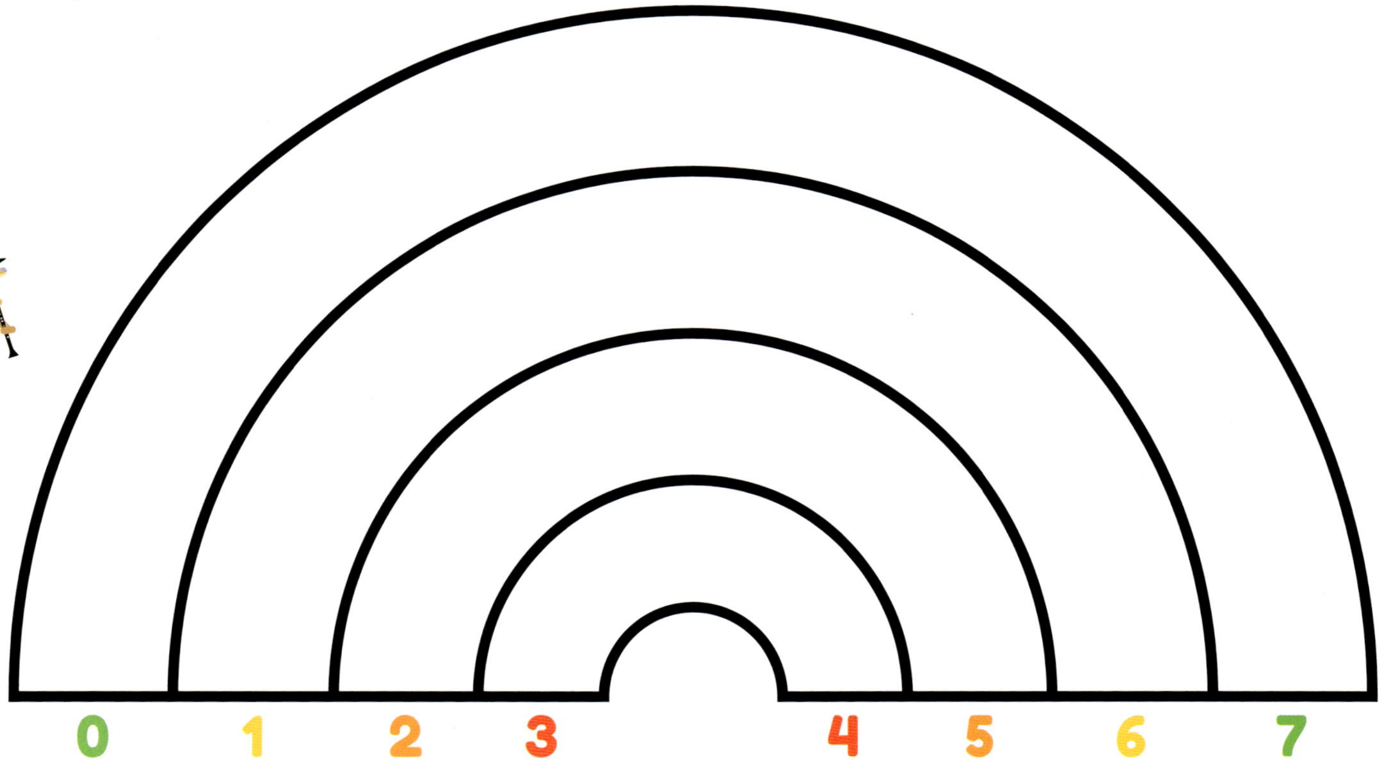

EIGHT 8

You can make the sum of 8 in different ways, using different addition sentences.

Color the rainbow to connect the numbers that equal 8.

Color each part of the rainbow the same color as the two numbers that equal 8.

0 + 8 = 8	8 + 0 = 8
1 + 7 = 8	7 + 1 = 8
2 + 6 = 8	6 + 2 = 8
3 + 5 = 8	5 + 3 = 8
4 + 4 = 8	

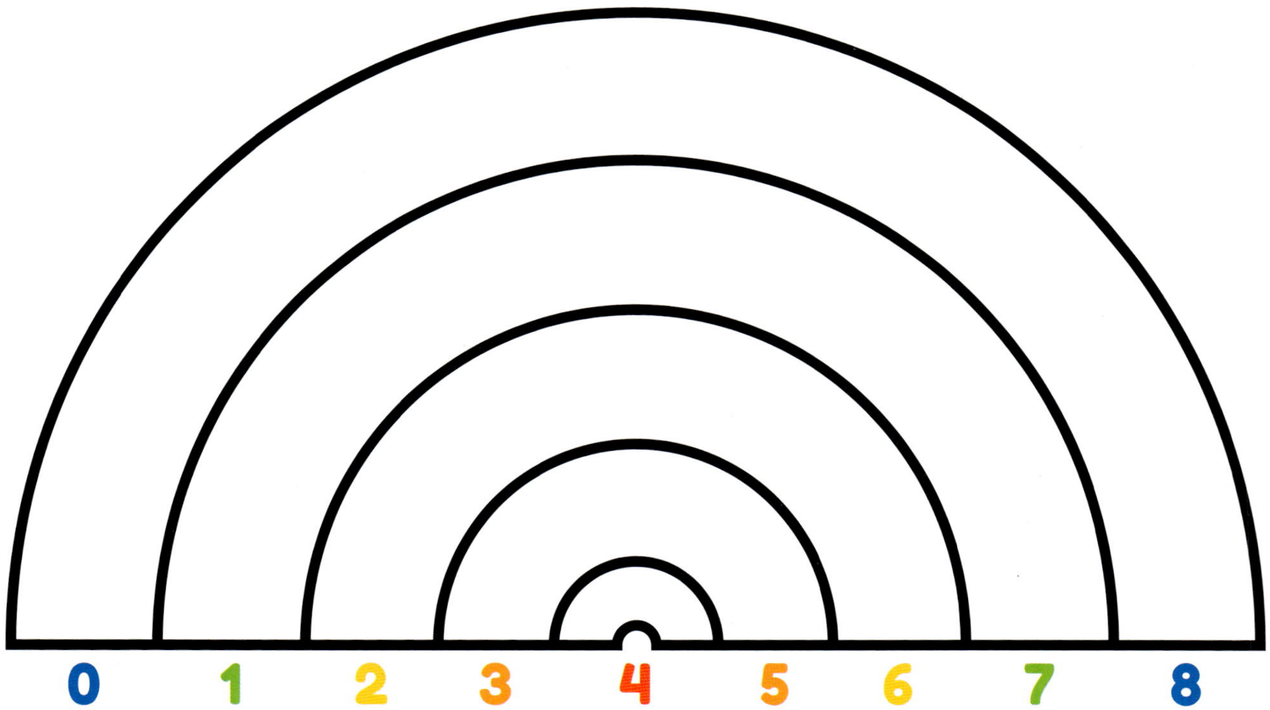

Notice that the number 4 is added to itself to equal the number 8.
4 + 4 = 8.

NINE 9

You can make the sum of 9 in different ways, using different addition sentences.

0 + 9 = 9	9 + 0 = 9
1 + 8 = 9	8 + 1 = 9
2 + 7 = 9	7 + 2 = 9
3 + 6 = 9	6 + 3 = 9
4 + 5 = 9	5 + 4 = 9

Color the rainbow to connect the numbers that equal 9.

Color each part of the rainbow the same color as the two numbers that equal 9.

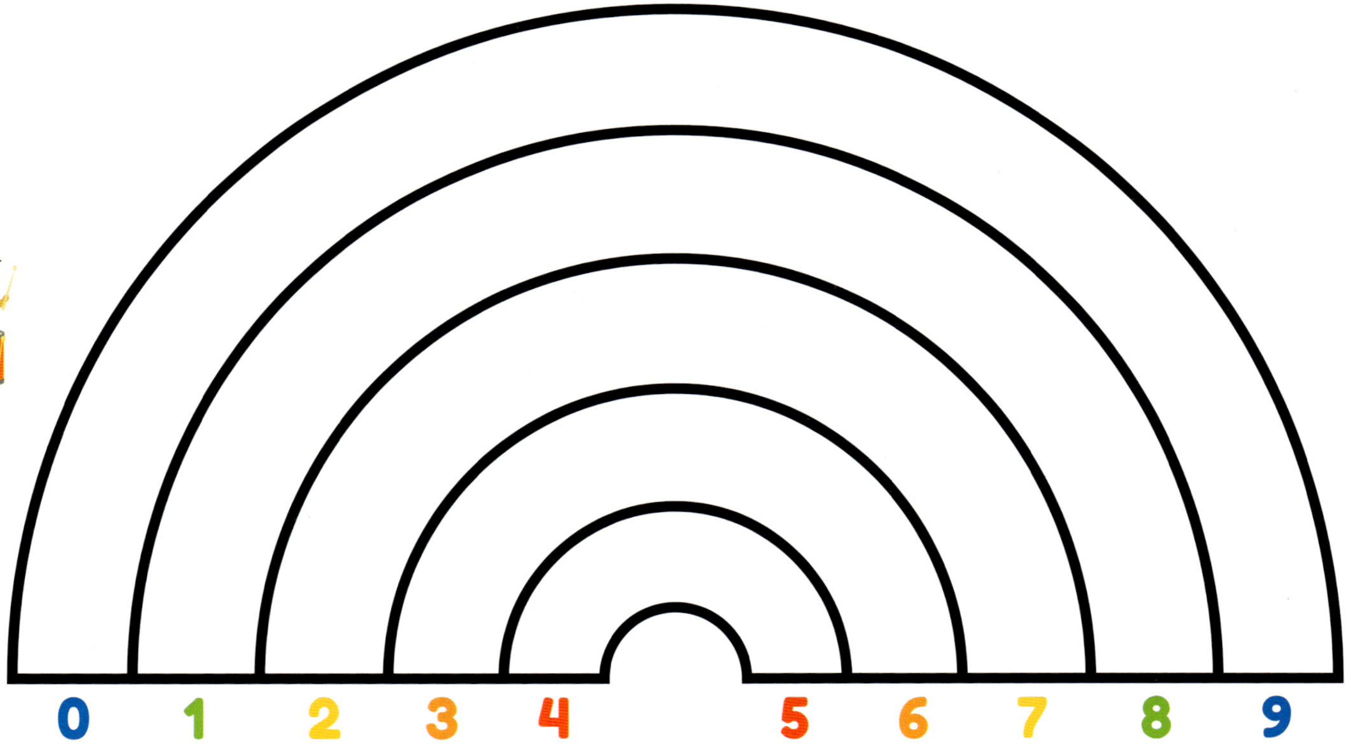

At Sixes and Sevens

Help the number 6 find its way through the maze. Color the sums that equal 6 **pink**. Color the sums that equal 7 **blue**.

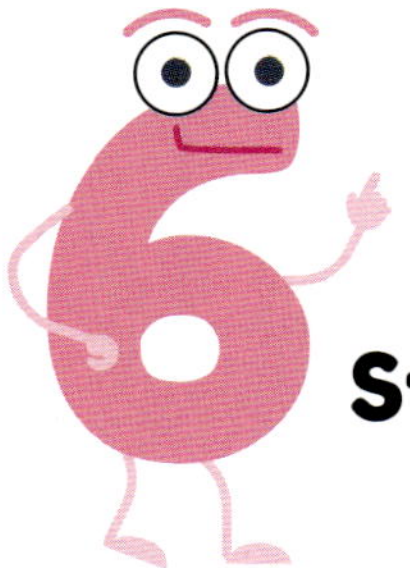

Start →

5 + 1	2 + 4	6 + 1	2 + 5	4 + 3	1 + 6
5 + 2	3 + 3	7 + 0	3 + 4	0 + 7	2 + 5
0 + 7	6 + 0	4 + 2	1 + 5	0 + 6	6 + 1
4 + 3	6 + 1	5 + 2	4 + 3	3 + 3	2 + 4

→ Finis

Puzzle Pieces

Fill in the missing number on each puzzle piece to finish each addition sentence. Then color the addition facts for the number 8 **orange**. Color the addition facts for number 9 **purple**. The first one has been done for you.

0 + | 8 | = 8

9 + | | = 9

3 + | | = 8

7 + | | = 9

| + 1 | = 8

| + 2 | = 9

| + 6 | = 8

4 + | | = 9

4 + | | = 8

Ways to Make 10

TEN 10

You can make the sum of 10 in different ways, using different addition sentences.

Color the rainbow to connect the numbers that equal 10.

Color each part of the rainbow the same color as the two numbers that equal 10.

0 + 10 = 10	10 + 0 = 10
1 + 9 = 10	9 + 1 = 10
2 + 8 = 10	8 + 2 = 10
3 + 7 = 10	7 + 3 = 10
4 + 6 = 10	6 + 4 = 10
5 + 5 = 10	

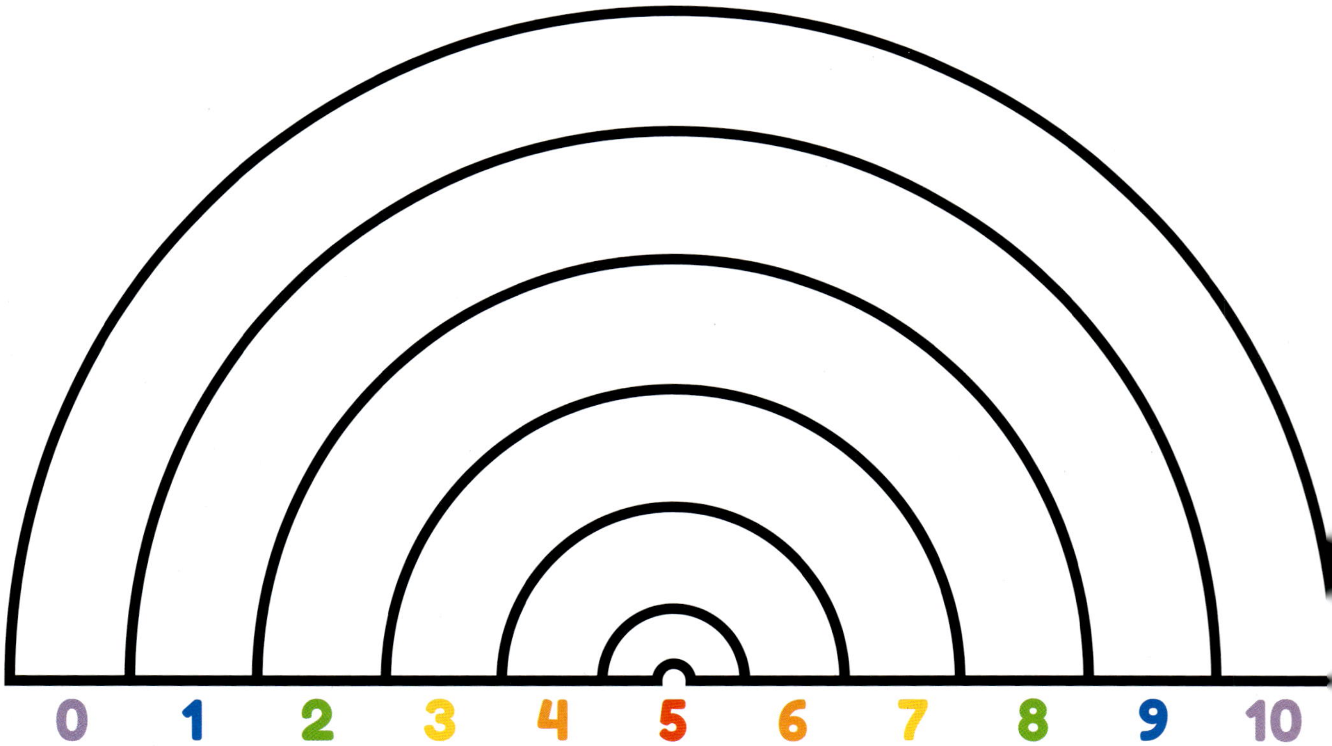

Notice that the number 5 is added to itself to equal the number 10.
5 + 5 = 10.

Ten Frames

There are other ways to show the number 10. One of them is called a *ten frame*.

Some ten frames have numbers.

1	2	3	4	5
6	7	8	9	10

Some ten frames have dots.

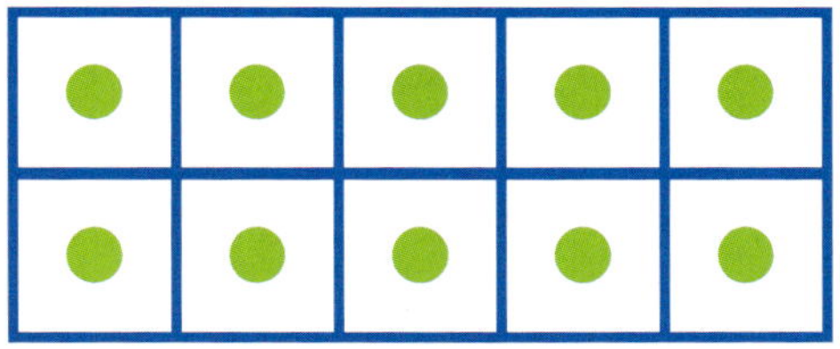

Some ten frames have pictures.

Some ten frames have shaded boxes.

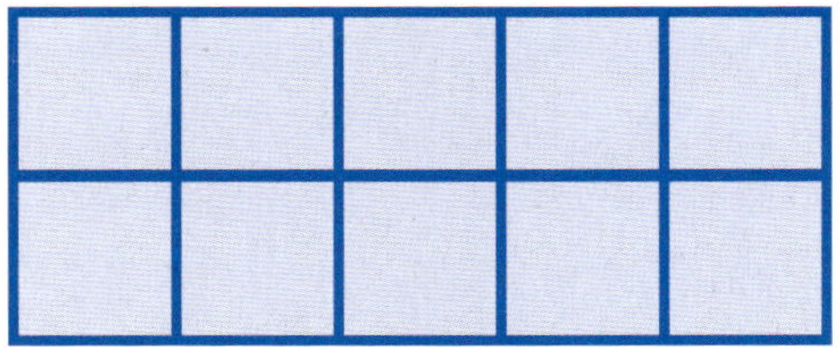

Ten frames can show any number up to 10. The number 1 is always shown in the top left corner of the ten frame. When you have a larger number, like 9, the boxes are filled from left to right and top to bottom.

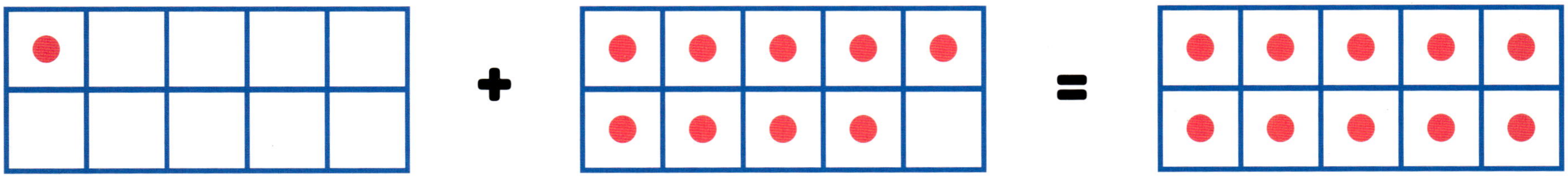

Discover

Let's Make Ten

- Count the green dots in the ten frame.
- Write the number in the first part of the addition sentence.
- Add enough dots to fill in the rest of the ten frame.
- Count the dots you added and write that number in the second part of the sentence.
- Write the sum after the equal sign.

  The first one has been done for you.

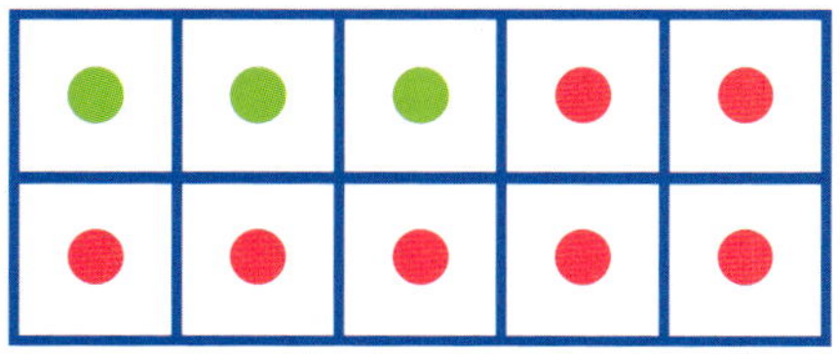

3 + 7 = 10

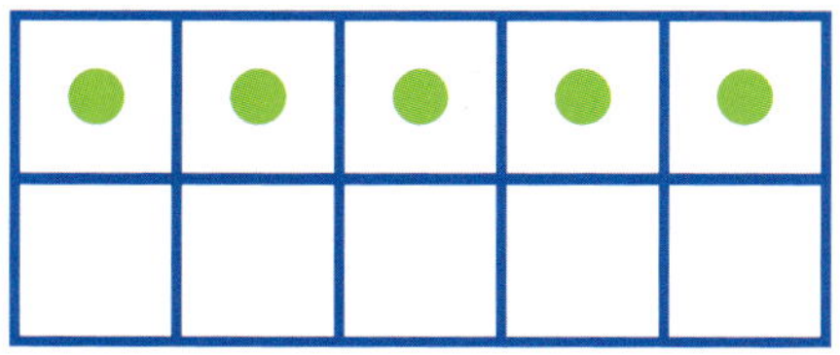

☐ + ☐ = ☐

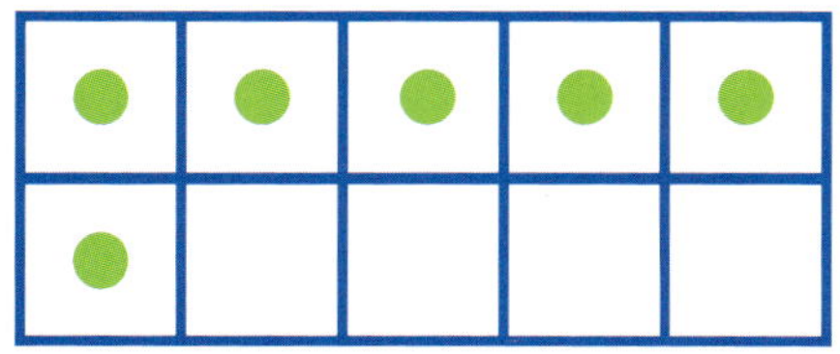

☐ + ☐ = ☐

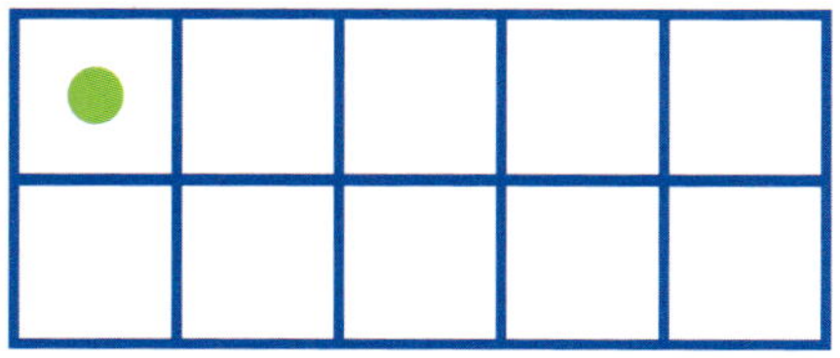

☐ + ☐ = ☐

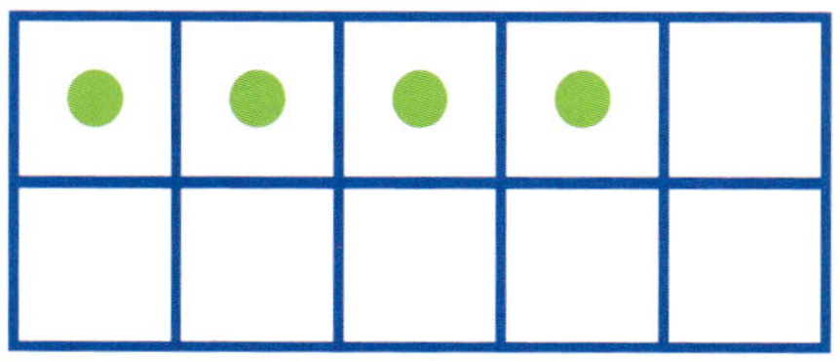

☐ + ☐ = ☐

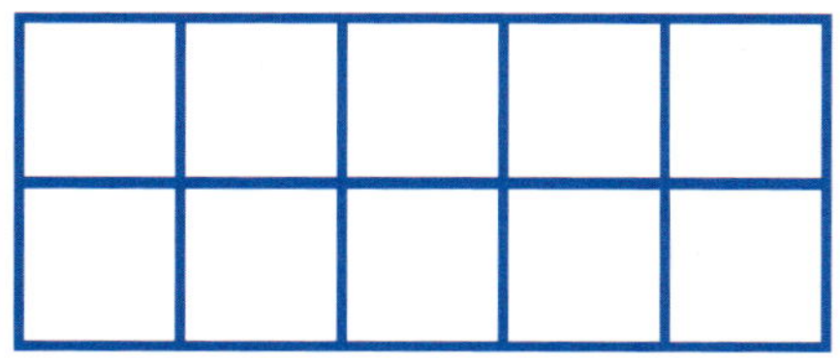

☐ + ☐ = ☐

Ten Pups

Read the storybook *Ten Pups* with your child. Every page contains a different number of pups, but the total number of pups always equals 10. There are also math sentences and ten frames on every page to provide different ways for your child to visualize the numbers in the story.

After reading, use these ideas to continue the learning:

- Play simple board games that require rolling dice to move around the board.
- Draw dots for the numbers 1-6 on pairs of index cards. Play a matching game, having your child call out the number when a card is flipped over.
- Draw dots for the numbers 1-9 on paper plates. Scatter the plates on the floor. Call out a number and have your child toss a beanbag or other soft item onto the matching plate.

The ability to "see" a small number of objects and know "how many" is an important skill called *subitizing*. As your child's number sense grows, the ability to know the amount of something without having to count it first will help with math fact fluency.

Launch

Vertical Addition

Up and Down

Vertical addition is a different way of adding. Instead of the numbers lining up in a row, they are lined up in a stacked pattern—with one number on top and the other number below.

The plus sign is placed to the left of the second number.

A line underneath the second number separates the numbers being added from the sum. It serves the same purpose as the equal sign. The sum is placed below the line.

$$\begin{array}{r} 2 \\ +\ 4 \\ \hline 6 \end{array}$$

Equal sign (the line)

Sum (6)

Horizontal and Vertical

Draw a line between the matching horizontal and vertical addition sentences.

5 + 5 = 10	4 + 2 = 6	1 + 8 = 9	5 + 2 = 7

$$\begin{array}{r} 4 \\ +\ 2 \\ \hline 6 \end{array} \qquad \begin{array}{r} 5 \\ +\ 5 \\ \hline 10 \end{array} \qquad \begin{array}{r} 5 \\ +\ 2 \\ \hline 7 \end{array} \qquad \begin{array}{r} 1 \\ +\ 8 \\ \hline 9 \end{array}$$

Add It Up

Add. Write the sum below the equal sign.

$$\begin{array}{r} 3 \\ +\,1 \\ \hline \end{array}$$

$$\begin{array}{r} 6 \\ +\,4 \\ \hline \end{array}$$

$$\begin{array}{r} 5 \\ +\,2 \\ \hline \end{array}$$

$$\begin{array}{r} 3 \\ +\,6 \\ \hline \end{array}$$

$$\begin{array}{r} 4 \\ +\,3 \\ \hline \end{array}$$

$$\begin{array}{r} 7 \\ +\,2 \\ \hline \end{array}$$

Spin-a-Sum

Take a Spin

Spin the spinner and write the number in the first box on the next page. Spin again and write the number in the second box. Then write the sum after the equal sign. Keep playing until all the boxes are filled.

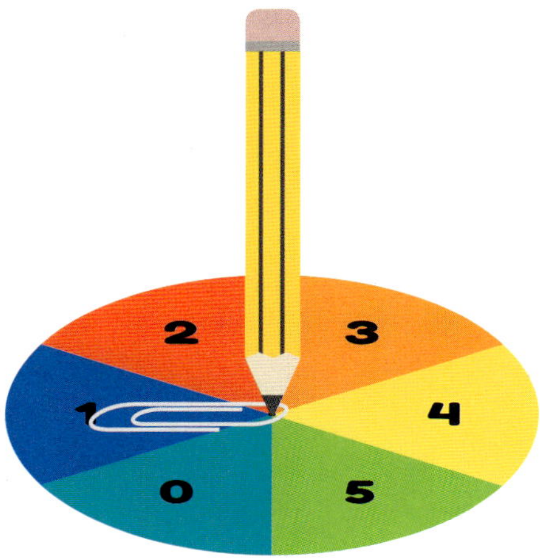

How to Make the Spinner: Place one end of the paper clip in the center of the circle. Put the pencil tip inside the paper clip on the center dot. Now flick the paper clip so it spins around the pencil.

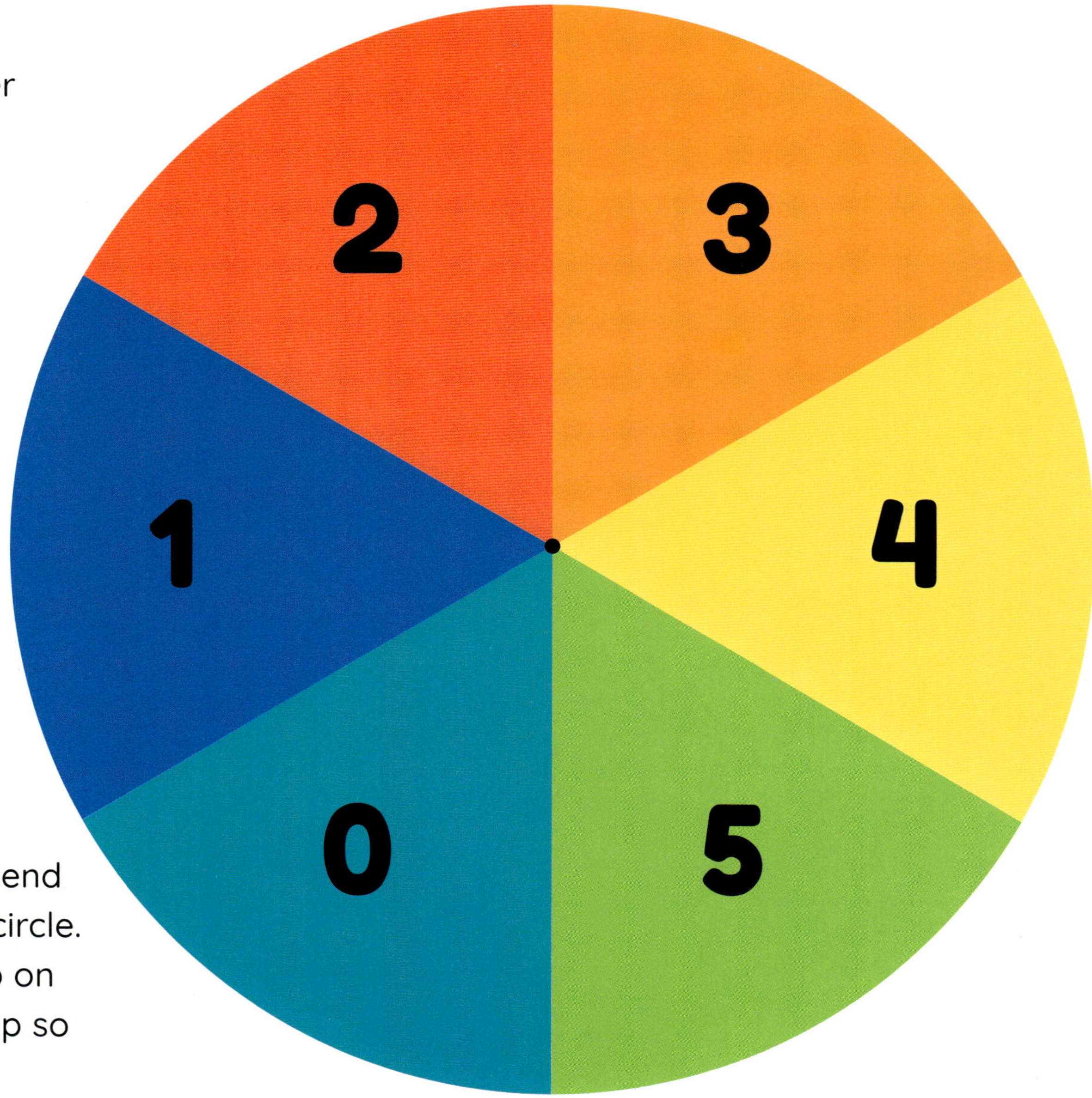

Materials needed: sharpened pencil, paper clip

□ + □ = □   □ + □ = □   □ + □ = □

□ + □ = □   □ + □ = □   □ + □ = □

□ + □ = □   □ + □ = □   □ + □ = □

Your child has completed Step 3!
Celebrate by placing a star sticker in the workbook.

HOOKED ON MATH

It's Time to Celebrate!

____

has completed Math Fundamentals Steps 1, 2, & 3 and is now ready to move on to the Math Fundamentals Steps 4 & 5 Workbook!

SKILLS MASTERED

- Counting 1 to 20
- Ordering numbers
- Adding to 10
- Number sentences
- Adding with 0
- Addition facts to 10
- Vertical addition

Don't forget to share your success and tag us on social media so we can celebrate with you!

Credits

Design Team
Julie Linden Frazier
Lindsey Frazier

Editorial Team
Donna Giachetti
Cynthia Stierle
Austin Feld

Illustrations
Dubit Limited

Educational Advisor
Lesley Freed, Curriculum Specialist

Published by Sandviks HOP, Inc.

Printed in China

HOOKED ON MATH

Math Fundamentals

18018 (05/22)

HOOKED ON MATH

Math Fundamentals

Steps 4 & 5

Workbook

Tips for Success

- Math Fundamentals introduces your child to the building blocks of math—number sense, addition, and subtraction.
- Work at a pace that's right for you and your child.
- Fit your session length to your child's attention span. It's okay to work with the App one day and this workbook the next.
- Before beginning, flip through the workbook and note any materials you might need. You should plan to have pencils, a pair of dice, and crayons ready.
- You'll see parent tips throughout this workbook. These tips are designed to answer questions you or your child might have about the learning.
- Repetition and review are essential to basic math fluency. Each workbook step builds on the ones before, so make sure your child has mastered the current lesson before moving forward.
- As you go through the workbook, encourage your child to explore the connection between numbers and the real world. Reinforce concepts with familiar household objects such as toys or snacks to make adding and subtracting more memorable.
- Have fun with it!

HOOKED ON MATH

Math Fundamentals

Steps 4 & 5

Workbook

This workbook and the included storybooks introduce fundamental math skills.

Step 4: Subtracting within 10
- Subtraction from 10
- Number sentences
- Subtracting with 0

Step 5: Subtraction Practice
- Counting back to subtract
- Subtraction facts to 10
- Vertical subtraction

Contents

Lesson Overview

Welcome! This workbook provides supplemental activities that will solidify and expand the learning your child is enjoying with the Hooked on Math App. Follow the easy, step-by-step instructions to guide your child's exploration of numbers and operations. Every lesson helps your child build a strong foundation in fundamental math skills while having fun!

1. Launch

Your child will be introduced to math concepts and skills through interactive lessons in the App on your mobile device or Windows or Mac computer. Download the App at: iOS

hookedonmath.com/get-the-app

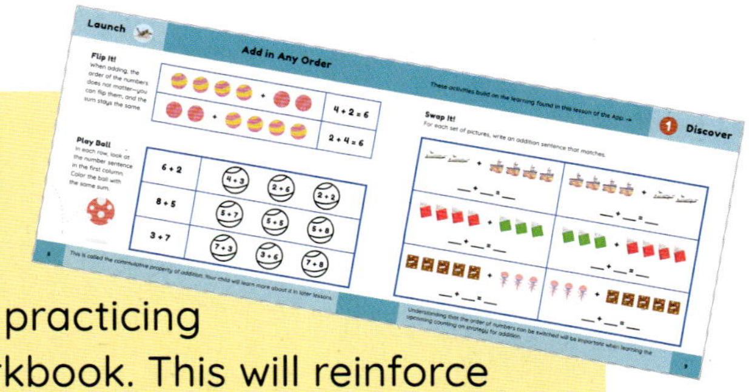

2. Discover

Next, your child will have fun exploring math concepts and practicing new skills in the workbook. This will reinforce the App lessons and provide a tangible application of the skills learned.

3. Play

In addition to the workbook activities, your child will enjoy two math-based storybooks that will strengthen their learning and add to the fun!

4. Celebrate

Celebrate your child's progress by placing a sticker in the workbook as learning milestones are reached. This also provides a helpful marker to remember where to begin next time.

Special notes are included in this space to provide additional context related to what is being taught.

All Aboard!

Load the train by writing the numbers in each row in the correct order from least to greatest. The first one has been done for you.

Numbers	Train
9 3 7	3 7 9
4 0 2	
8 3 6	
1 5 2	

Make It 10!

Write the different ways to make 10 in the spaces below. Use the rainbow to help you remember which friends make 10. The first one has been done for you. **Flip tip:** Don't forget that you can switch the order of the numbers being added together.

0 + 10 = 10	___ + ___ = 10	___ + ___ = 10
___ + ___ = 10	___ + ___ = 10	___ + ___ = 10
___ + ___ = 10	___ + ___ = 10	___ + ___ = 10
___ + ___ = 10		___ + ___ = 10

Roll a Fact Family

Roll a pair of dice. Count all the dots you rolled and write that number in the top corner of the triangle. Next, write the numbers from each die in the two lower corners. The first one has been done for you.

Keep rolling until all of the triangles are filled. If you roll the same fact family twice, roll again.

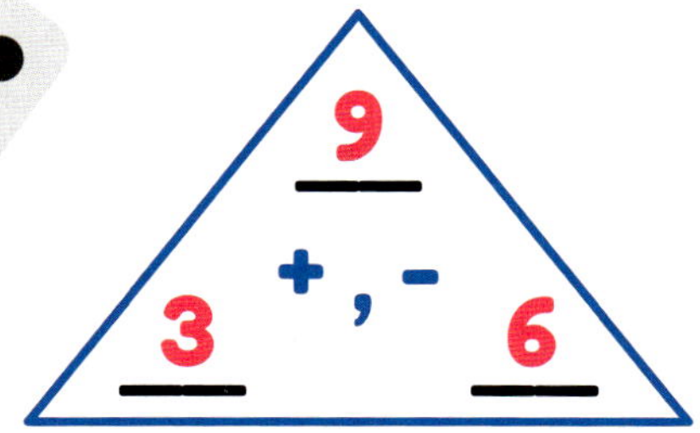

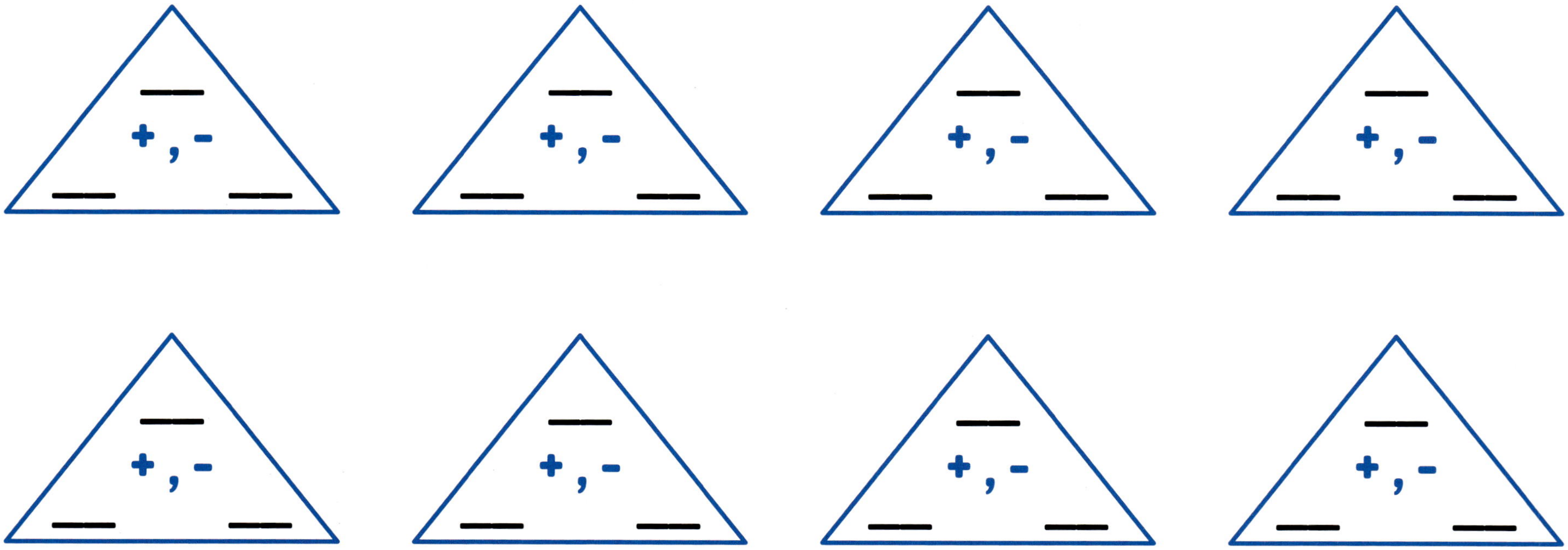

Materials needed: Pair of dice

Adding with 0 — Review

Cookie Time!

Write the sum below each cookie. Color the cookies according to the Color Key. The first one has been done for you.

0 + 1 = 1

0 + 6 = ☐

0 + 3 = ☐

Color Key

- 1 = yellow
- 2 = orange
- 3 = green
- 4 = pink
- 5 = purple
- 6 = blue

0 + 4 = ☐

0 + 5 = ☐

0 + 2 = ☐

Materials needed: crayons or colored pencils in yellow, orange, green, pink, purple, and blue.
Celebrate finishing the review by placing a star sticker in the workbook.

Introducing Subtraction

Taking away a number from another number is called *subtraction*.

Story Time

Read the subtraction stories below. Count the items and write the answer in the box next to each story. The first one has been done for you.

There are 8 fish swimming in the ocean. After 4 fish swim away, how many fish are left?

4

There are 7 cars in the parking lot. Then 4 cars drive away. How many cars are left?

There are 6 ants at the picnic. Only 1 ant leaves with a sandwich. How many ants did not get a sandwich?

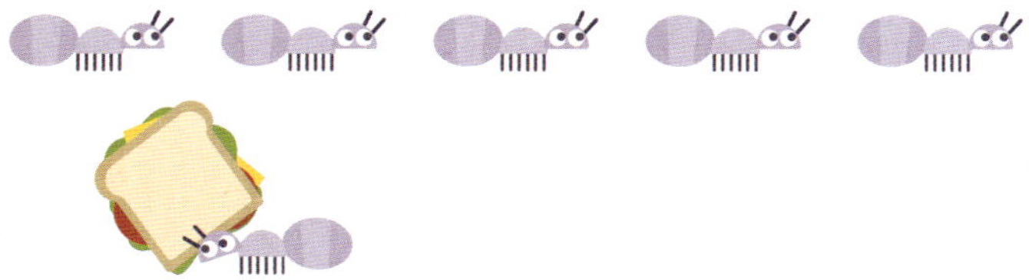

There are 10 apples in the basket. A hungry meerkat eats the 2 green apples. How many red apples are left?

Color the Story

Read the subtraction stories below. Color the pictures to match each story. Write the answer in the box. The first one has been done for you.

Tali's hat has 6 feathers. She gives 2 blue feathers to Zola and keeps the others, which are green. How many feathers does Tali have?

4

Chow has 9 balloons. He gives 3 red balloons to Kazoo. The rest of the balloons are yellow. How many yellow balloons does Chow have?

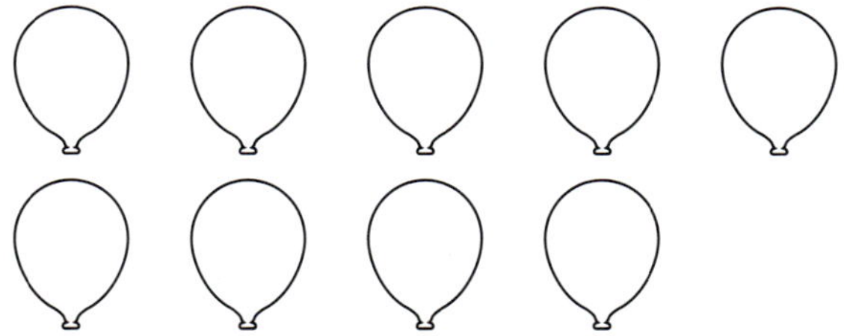

All 5 meerkats have umbrellas. There are 2 meerkats with orange umbrellas. A big wind blows away all the other umbrellas, which are purple. How many umbrellas were blown away?

Batu has 7 ice cream cones. He shares 2 pink cones with his pals. He eats the green ones since he loves mint chip! How many ice cream cones did Batu eat?

Understanding Subtraction

The symbol for subtraction is called the *minus sign* (-). It tells you to take away one number from another to find out how many are left.

Submarine Subtraction

You can use pictures to solve subtraction problems by crossing out. Look at each group of submarines on the left and match them to the correct subtraction problem on the right. The first one has been done for you.

	8 - 3
	6 - 2
	10 - 5
Draw your own subtraction picture here.	___ - ___

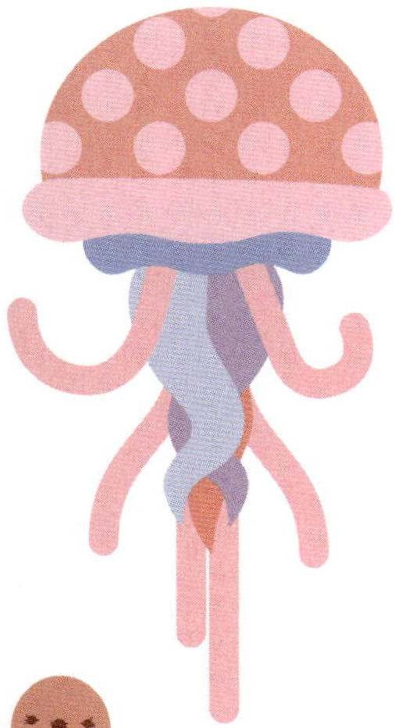

Linked Together

Roll a pair of dice. Count all the dots you rolled and write that number in the top box. Next, write the numbers from each die in the lower boxes. Make two subtraction sentences. Write the number from the top box in the first space. Write the other numbers in the remaining spaces.

10 - 6 = 4

10 - 4 = 6

___ - ___ = ___ ___ - ___ = ___ ___ - ___ = ___ ___ - ___ = ___

___ - ___ = ___ ___ - ___ = ___ ___ - ___ = ___ ___ - ___ = ___

Materials needed: Pair of dice

Math Talk

A *subtraction sentence* shows one number being taken away from another number to find the answer, or *difference*.

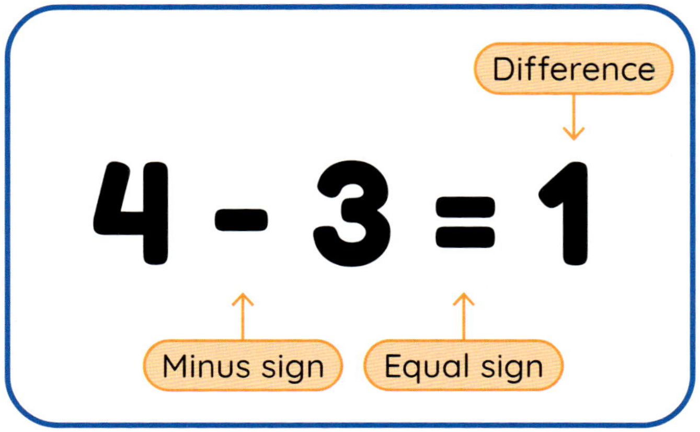

The signs in a math sentence tell you what mathematical action, or *operation*, you need to do.

- The *minus sign* tells you to take the second number away from the first number.
- The *equal sign* tells you that the quantity on one side of the sign is the same as on the other side.

See the Signs

Look at the math sentences below. Circle the largest number. Draw a box around the number that's being taken away. Underline the difference. Color the minus signs **blue**. Color the equal signs **red**. The first one has been done for you.

$5 - 4 = 1$ $\quad 4 - 2 = 2$ $\quad 9 - 2 = 7$

$8 - 5 = 3$ $\quad 10 - 6 = 4$ $\quad 7 - 2 = 5$

She Sees Shells

Look at the pictures and write the subtraction sentences.
Then write the difference. The first one has been done for you.

	Write the sentence	Difference
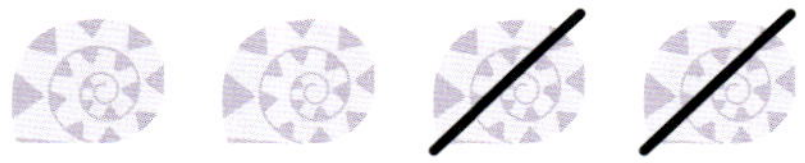	4 - 2 =	2
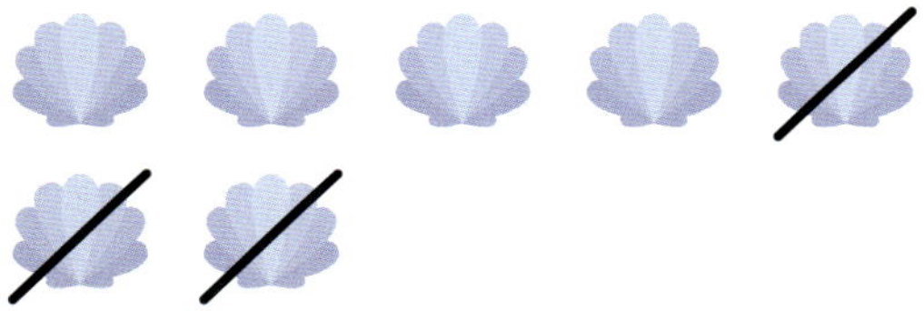	___ - ___ =	___
	___ - ___ =	___
	___ - ___ =	___

Subtracting with 0

Remember, just like in addition, the number 0 represents nothing. When taking away 0 from any number, it does not change the number.

A Hero Called Zero

The number 0 has a very important job. It is used as a placeholder. Without 0, you would not know the difference between the numbers 1, 10, and 100.

$$1 - 0 = 1$$

In the Right Place

Circle the correct answer to each subtraction sentence in the first column. The first one has been done for you.

5 - 0 =	50	500	(5)
12 - 0 =	120	102	12
15 - 0 =	150	15	105
8 - 0 =	8	800	80

Crack Up Code

Write the answer to each number sentence in the chart. Then write the letter from each box in the matching number space to answer the riddle.

s	e	i
9 - 0 = ___	3 - 0 = ___	2 - 0 = ___
o	**n**	**l**
6 - 0 = ___	1 - 0 = ___	7 - 0 = ___
m	**r**	**a**
4 - 0 = ___	5 - 0 = ___	8 - 0 = ___

Why did the Captain bring snails on the ship?

Because the Captain needed

___ ___ ___ ___   ___ ___ ___ ___ ___ ___ ___ ___!

9 6 4 3   9 1 8 2 7 6 5 9

You know that a *fact family* is made up of 3 numbers that are related to each other.

When the 2 smaller numbers are added together, they equal the largest number.	6 + 4 = 10 4 + 6 = 10
When 1 of the smaller numbers is taken away from the largest number, it equals the other smaller number.	10 - 4 = 6 10 - 6 = 4

10
+ , -
4 6

Fact Family Find

Look at each fact family.

- Circle the two smaller numbers in each family.
- Draw a box around the largest number in each family.
- Then write two addition and two subtraction sentences for each family.

10
+ , -
3 7

___ + ___ = ___
___ + ___ = ___
___ - ___ = ___
___ - ___ = ___

9
+ , -
6 3

___ + ___ = ___
___ + ___ = ___
___ - ___ = ___
___ - ___ = ___

Map Maze

Find and color the subtraction sentences that match the fact family subtraction sentences on the opposite page to find the way through the maze.

→ Start

10 - 4	9 - 6	9 - 3	5 - 4	9 - 4	7 - 3
10 - 5	5 - 2	10 - 6	10 - 7	5 - 3	5 - 4
8 - 1	4 - 0	4 - 2	9 - 6	9 - 3	8 - 6
7 - 6	5 - 1	9 - 1	10 - 0	10 - 3	10 - 4

→ Finish

Tip: Remind your child to look at the sentences, not the answers.

Salad Subtraction

Batu loves to eat vegetables. Read each story. Then cross out the number of vegetables Batu ate in each picture. Count what is left and write the subtraction sentence. The first one has been done for you.

There were 8 beets. Batu ate 6 of them.
How many beets are left?

8 - 6 = 2

There were 5 onions. Batu ate 4 of them.
How many onions are left?

___ - ___ = ___

There were 10 cabbages. Batu ate 3 of them.
How many cabbages are left?

___ - ___ = ___

There were 9 carrots. Batu ate 5 of them.
How many carrots are left?

___ - ___ = ___

At this stage in your child's math education, they are only learning to subtract positive numbers. Remind your child that a subtraction sentence must start with the largest number.

Color Key

1 = orange
2 = green
3 = red
4 = brown
5 = blue
6 = yellow
7 = purple
8 = dark green

Color by Number Subtraction

Find the difference between the numbers in each spot. Then use the key to color the picture.

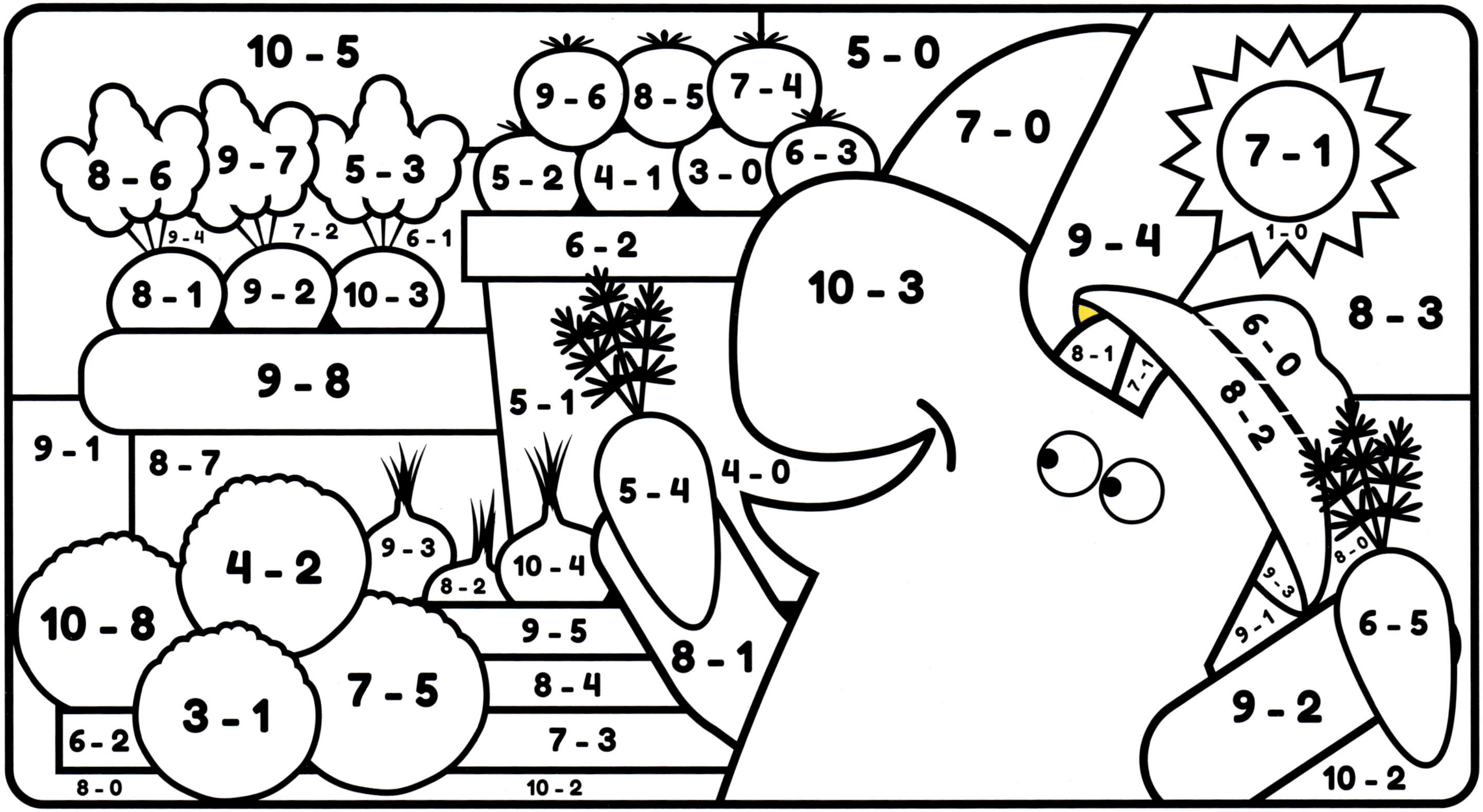

Materials needed: crayons or colored pencils in orange, green, red, brown, blue, yellow, purple, and dark green

Chow the Magician

Read the storybook *Chow the Magician* with your child. In this book, you will see different fact families. Learning these facts now will help your child solve equations with larger numbers later.

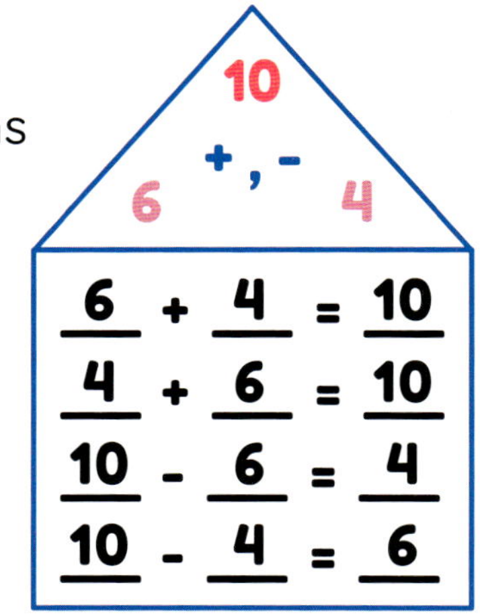

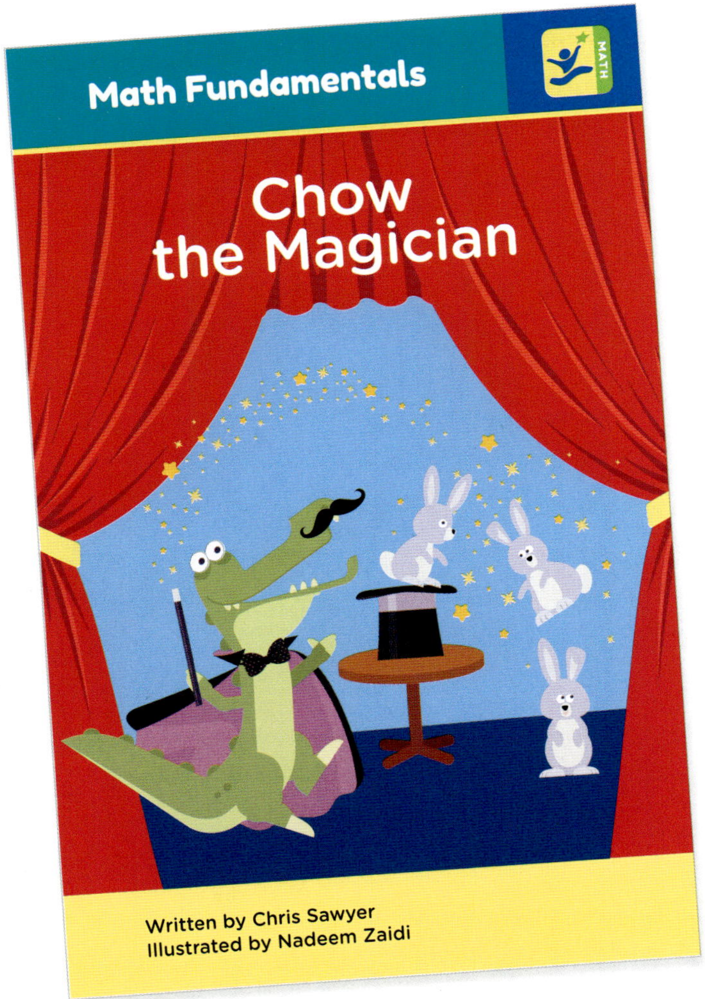

During reading:

- Encourage your child to study each fact family—especially the addition and subtraction sentences that correspond to the triangle.
- Periodically, cover up the math sentences and ask your child to tell you what math sentences make up the fact family, based on any two numbers in the triangle.

After reading:

- Ask your child questions about the book: which magic trick was their favorite? What tricks would they like to perform?
- Was Chow dreaming, or did his magic act really happen in the night?

Have a Ball!

Use the Color Key to write in the numbers in the matching spots so that each fact family can have a ball! The first one has been done for you.

Color Key

1 = orange
3 = green
4 = pink
6 = purple
7 = blue
10 = yellow

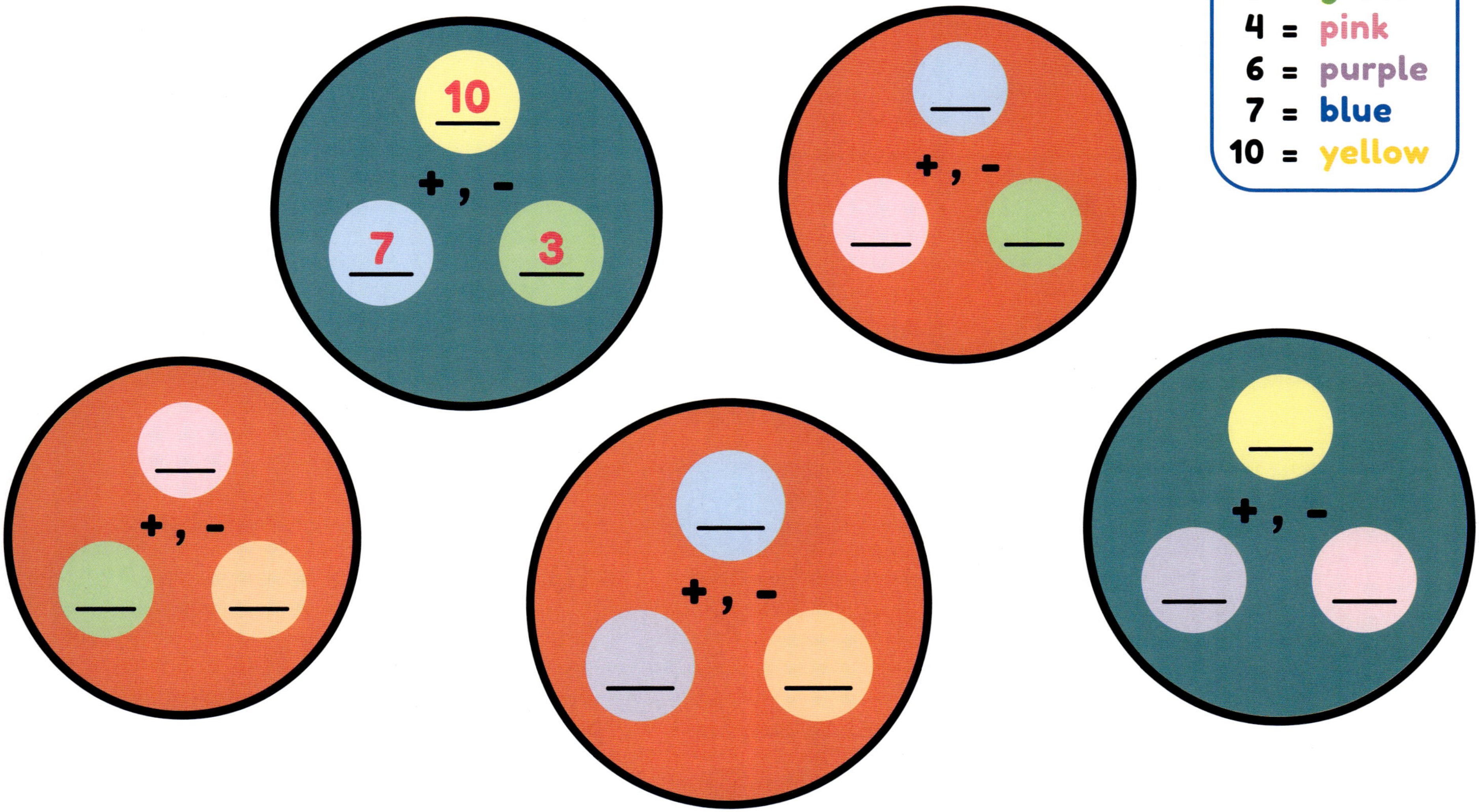

Your child has completed Step 4!
Celebrate by placing a star sticker in the workbook.

Launch 

Counting Back to Subtract

When the number being subtracted is small, you can use the *counting back strategy* to find the difference. A number line can help you!

Start at the greater number and count back the number you want to subtract. For example, to find the difference of 9 - 3, start at the number 9 and jump back 3 spaces to stop at number 6.

> The number line is a straight line that has numbers placed at equal spaces or segments along its length.

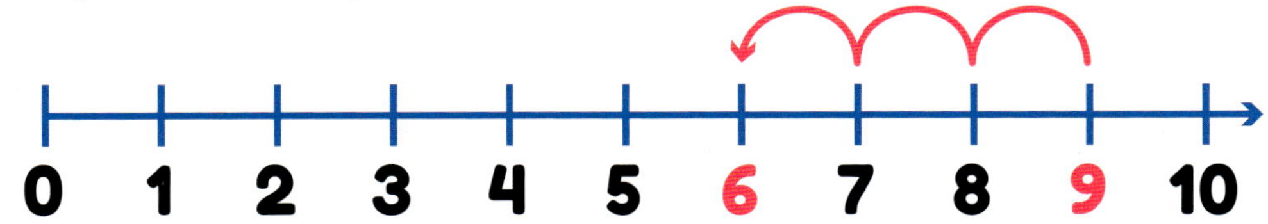

9 - 3 = 6

Line Up

Use the number line to solve the subtraction problem below.
Circle the starting number. Draw the jumps. Underline the difference.
The first one has been done for you.

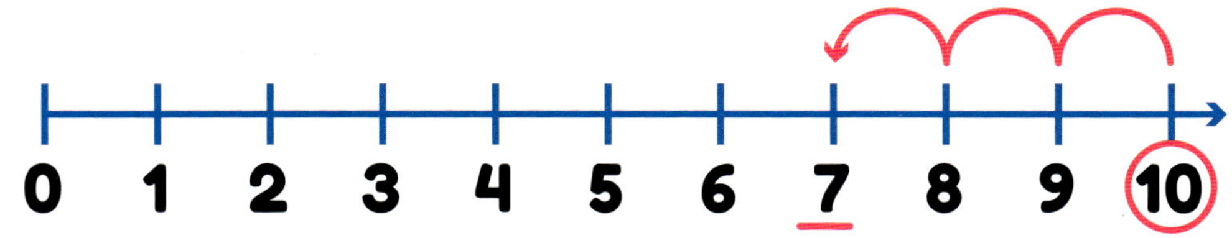

10 - 3 = 7

8 - 4 = ___

A number line is a useful tool that can help your child "see" subtraction in action. It is important that your child understands the concept of subtraction in order to build fluency with math facts.

Frog Friends

Help each frog reach its lily pad. Starting at the first number in the sentence, draw a jump back to the next lily pad. Keep jumping back the number you need to subtract. Circle the last lily pad and write the answer to finish the sentence.

9 - 3 = 6

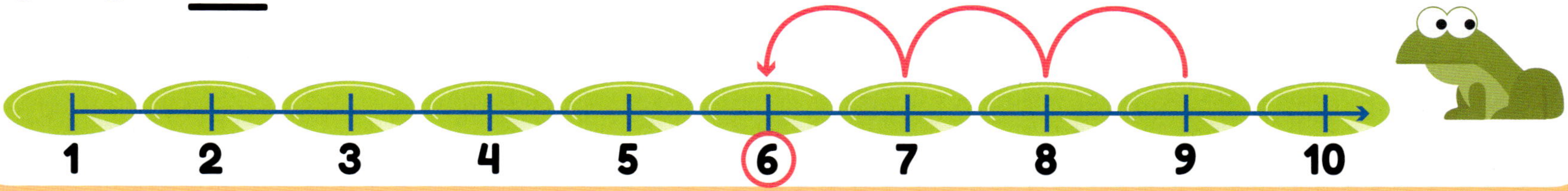

7 - 2 = ___

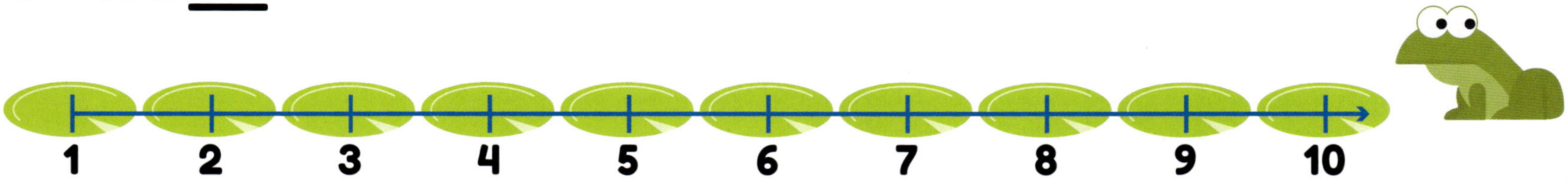

10 - 5 = ___

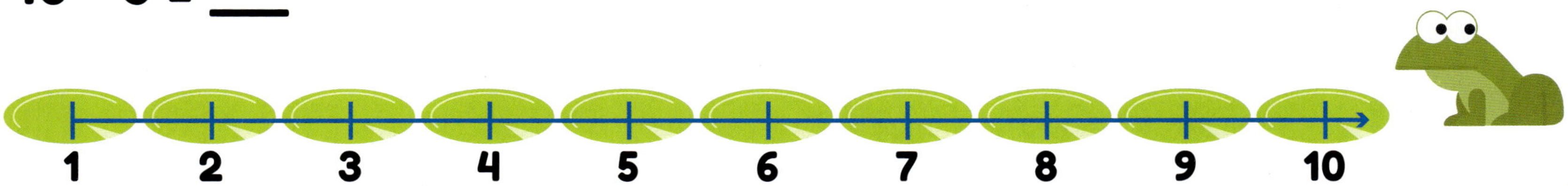

Subtract Lesser from Greater

In a subtraction sentence, the first number must be greater than the second number.	 is greater than  **6 > 4**
Unlike an addition sentence, the two numbers in a subtraction sentence cannot be flipped.	 **6 - 4**  **4 - 6**

Read each story. Write the subtraction sentence. Then check Yes or No in the box next to the story to answer the question. The first one has been done for you.

	Sentence	Yes	No
Kazoo picked 5 oranges. Then 7 meerkats arrived. Can Kazoo give each meerkat an orange?	5 - 7		✓
Tali has just baked 6 muffins when 3 friends come to see her. Can Tali give each friend a muffin?			
Chow has 3 nails. He wants to hang 5 pictures. Can Chow hang all the pictures?			

Subtracting to reduce an amount to a negative number is a concept that will be introduced later. For now, it is important for your child to master subtracting positive numbers.

Which Picture?

Which picture correctly shows the subtraction sentence in the middle column?

- In each row, circle the picture that matches the subtraction sentence.
- Draw an **X** through the pictures that do not match any sentences. The first one has been done for you.

– 	**10 – 8**	– 
– 	**5 – 1**	– 
– 	**7 – 6**	– 

Subtraction Practice

It's Bananas!

Look at each bunch of bananas. Cross out the number of bananas needed to match the subtraction sentence.

3 - 2 = 1

5 - 3 = 2

7 - 5 = 2

10 - 6 = 4

8 - 4 = 4

9 - 8 = 1

In subtraction, the first (or greater) number is called the *minuend*. The second (or lesser) number is called the *subtrahend*.

See It! Write It! Solve It!

Look at the pictures. Write each part of the subtraction sentence. The first one has been done for you.

See It!	Write It!	Solve It!
	4 - 2 =	2
	___ - ___ =	___
	___ - ___ =	___
	___ - ___ =	___

Subtraction Picture Equations

You can use many different visual tools in math. They help you see more clearly how to figure out a math equation.

Cap It Off

Look at each subtraction picture equation. Circle the picture that shows the correct difference. The first one has been done for you.

Which is the correct difference?

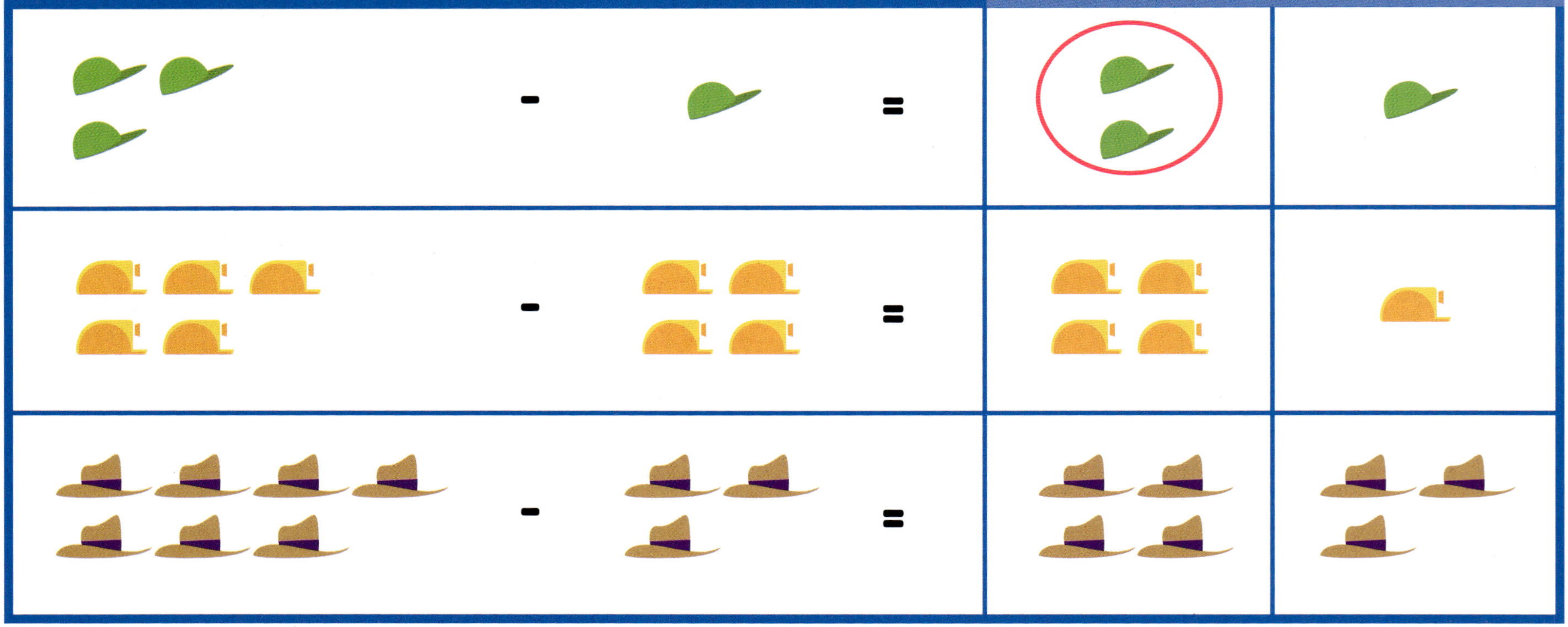

Draw the Story

Read the subtraction stories below. For each story, draw the largest number of items in the box. Cross out the number of items that are being taken away. Circle the difference. The first one has been done for you.

Chow has 5 kites. There are 3 meerkats. Chow gives each one a kite. How many kites does Chow have left?

There are 8 plates. If 5 pals are having lunch, how many extra plates are there?

Zola borrows 6 books from the library. She reads and returns 3 of them. How many books does Zola have left to read?

Batu has 10 pears. He eats 9 of them. How many pears are left?

Vertical Subtraction

Up and Down

Vertical subtraction is a different way to show a subtraction equation. Instead of the numbers lining up in a row, they are stacked—with the greater number on top and the lesser number below.

The minus sign is placed to the left of the second number.

A line underneath the second number serves the same purpose as the equal sign. The difference is placed below the line.

$$\begin{array}{r} 9 \\ -\ 4 \\ \hline 5 \end{array}$$

Equal sign

Difference

Find the Difference

Look at the pictures in each box below and write the difference.

$$\begin{array}{r} 6 \\ -\ 2 \\ \hline \end{array}$$

$$\begin{array}{r} 8 \\ -\ 6 \\ \hline \end{array}$$

$$\begin{array}{r} 4 \\ -\ 1 \\ \hline \end{array}$$

$$\begin{array}{r} 9 \\ -\ 3 \\ \hline \end{array}$$

Line It Up

Match each subtraction sentence on the left to a picture. Then match the picture to the vertical subtraction sentence on the right.

Across	Picture	Vertical
8 - 2		6 - 3
10 - 5		7 - 4
6 - 3		8 - 2
7 - 4	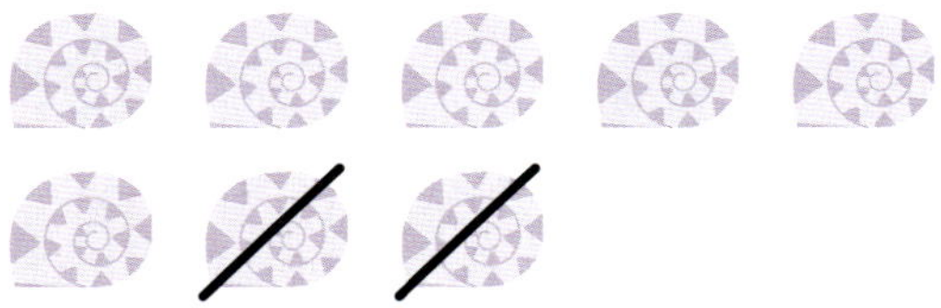	10 - 5

Hop Back

Help each bunny through the carrot patch. Starting at the first number in the sentence, draw a jump back to the next carrot. Keep jumping back the number you need to subtract. Circle the last carrot and write the answer to finish the sentence.

9 - 3 = ___

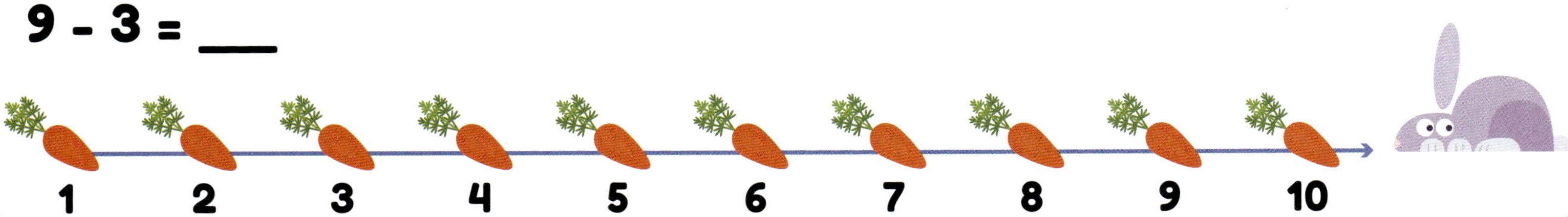

8 - 6 = ___

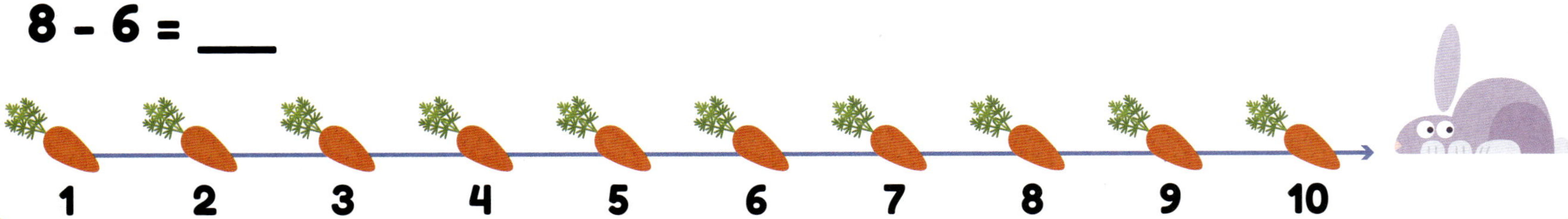

10 - 9 = ___

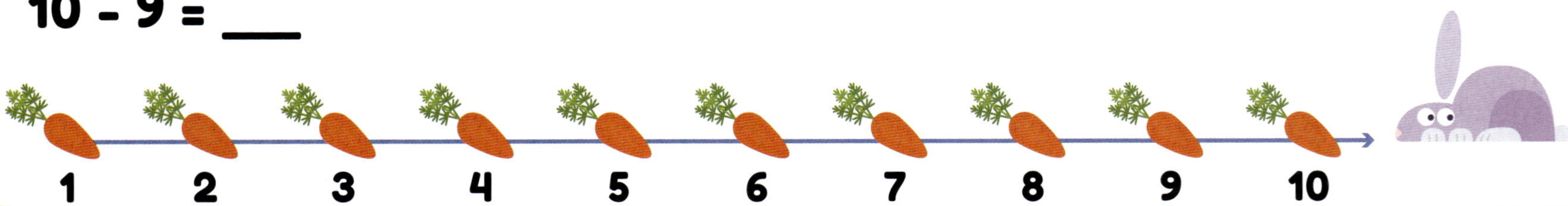

Tic-Tac-Toe

Look at each subtraction sentence.

- Draw an **O** around the sentences where the greater number is the first number.
- Draw an **X** through the sentences where the lesser number is first.
- See if the greater or lesser numbers win in each game.

Riddle: Why was 6 afraid of 7?

Answer: Because 7-8-9.

0 - 5	8 - 3	4 - 8
3 - 2	6 - 4	3 - 4
2 - 9	4 - 0	10 - 7

1 - 8	5 - 8	3 - 1
8 - 4	6 - 3	2 - 9
9 - 6	1 - 2	5 - 3

The Alien Picnic

Read the storybook *The Alien Picnic* with your child. In this book, you will come upon many number lines. These offer a reminder of how helpful number lines can be when performing simple subtraction equations.

During reading:

- Encourage your child to examine the number lines at the bottom of the pages. They can use a finger to "hop" backward along the number line, saying each number aloud as they go.
- This is a good chance to confirm your child's understanding of the idea that a subtraction sentence must always start with the largest number.

After reading:

- Prepare a picnic with your child! It can take place indoors or outdoors. Have them count aloud as they pack the foods into your picnic tote.
- Ask your child if they would have agreed to go up in the spaceship with the aliens at the end of the story. What do they think the alien's home planet looks like? Ask them to draw a picture of an alien's house.

Celebrate your child's progress by placing a star sticker in the workbook.

Subtraction Action

Fill in the missing numbers by finishing the subtraction sentences.

				8	-	1	=	
				-				-
10	-	7	=					6
-		-		=				=
6					-	4	=	
=		=				-		
	-	2	=			1		
						=		
		6	-	3	=			

Your child has completed Step 5!
Celebrate by placing a star sticker in the workbook.

HOOKED ON MATH

It's Time to Celebrate!

has completed Math Fundamentals Steps 4 & 5
and is now ready to move on to the
Math Fundamentals Steps 6 & 7 Workbook!

SKILLS MASTERED

- Subtraction from 10
- Number sentences
- Subtracting with 0
- Counting back to subtract
- Subtraction facts to 10
- Vertical subtraction

Don't forget to share your success and tag us on social media so we can celebrate with you!

Credits

Design Team
Julie Linden Frazier
Lindsey Frazier

Editorial Team
Donna Giachetti
Cynthia Stierle
Austin Feld

Illustrations
Dubit Limited

Educational Advisor
Lesley Freed, Curriculum Specialist

Published by Sandviks HOP, Inc.

Printed in China

HOOKED ON MATH

Math Fundamentals

18021 (05/22)

HOOKED ON MATH

Math Fundamentals

Steps 6 & 7

Workbook

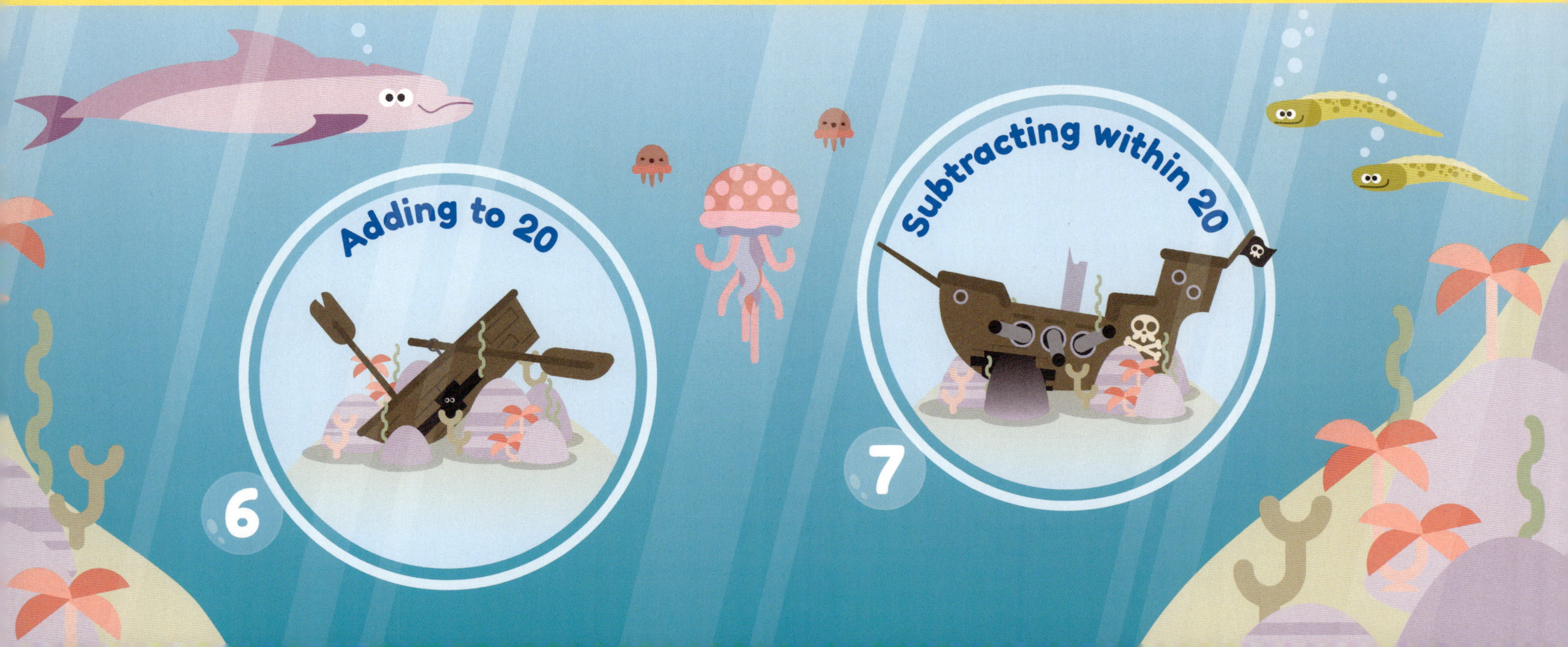

Tips for Success

- Math Fundamentals introduces your child to the building blocks of math—number sense, addition, and subtraction.
- Work at a pace that's right for you and your child.
- Fit your session length to your child's attention span. It's okay to work with the App one day and this workbook the next.
- Before beginning, flip through the workbook and note any materials you might need. You should plan to have pencils, a pair of dice, and crayons ready.
- You'll see parent tips throughout this workbook. These tips are designed to answer questions you or your child might have about the learning.
- Repetition and review are essential to basic math fluency. Each workbook step builds on the ones before, so make sure your child has mastered the current lesson before moving forward.
- As you go through the workbook, encourage your child to explore the connection between numbers and the real world. Reinforce concepts with familiar household objects such as toys or snacks to make adding and subtracting more memorable.
- Have fun with it!

Math Fundamentals

Steps 6 & 7

Workbook

This workbook and the included storybooks introduce fundamental math skills.

Step 6: Adding to 20
- Add in any order
- Counting on
- Doubles + 1

Step 7: Subtracting within 20
- Counting back
- Part-part-whole
- Missing doubles
- Finding missing numbers

Contents

Lesson Overview

Welcome! This workbook provides supplemental activities that will solidify and expand the learning your child is enjoying with the Hooked on Math App. Follow the easy, step-by-step instructions to guide your child's exploration of numbers and operations. Every lesson helps your child build a strong foundation in fundamental math skills while having fun!

1. Launch

Your child will be introduced to math concepts and skills through interactive lessons in the App on your mobile device or Windows or Mac computer. Download the App at: **hookedonmath.com/get-the-app**

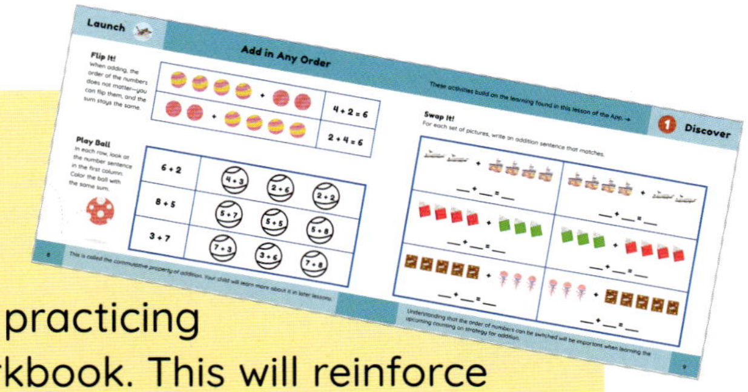

2. Discover

Next, your child will have fun exploring math concepts and practicing new skills in the workbook. This will reinforce the App lessons and provide a tangible application of the skills learned.

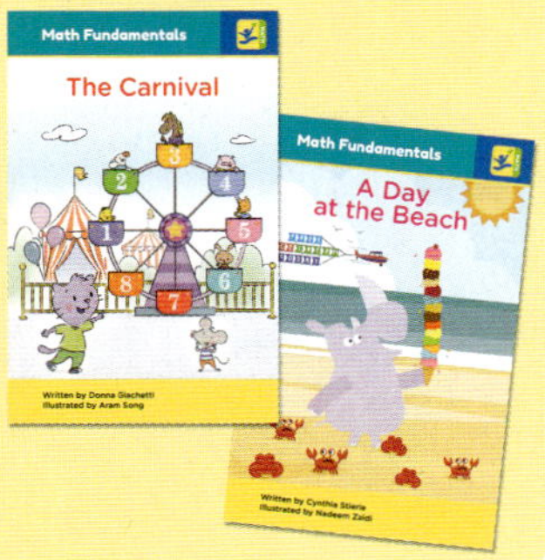

3. Play

In addition to the workbook activities, your child will enjoy two math-based storybooks that will strengthen their learning and add to the fun!

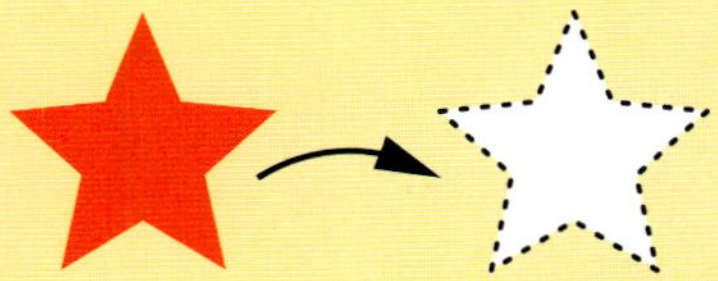

4. Celebrate

Celebrate your child's progress by placing a sticker in the workbook as learning milestones are reached. This also provides a helpful marker to remember where to begin next time.

Special notes are included in this space to provide additional context related to what is being taught.

Birthday Surprise!

Read the subtraction stories. Count the party items and write the answers to the questions.

Math Fruit Salad

Read each story. Cross out the number of fruits that Batu puts into the bowl. Count how many are left. Write the subtraction sentence.

There are 6 bananas. Batu uses 5. How many bananas are left?

___ - ___ = ___

There are 10 apples. Batu uses 8. How many apples are left?

___ - ___ = ___

There are 7 pears. Batu uses 3. How many pears are left?

___ - ___ = ___

There are 9 oranges. Batu uses 4. How many oranges are left?

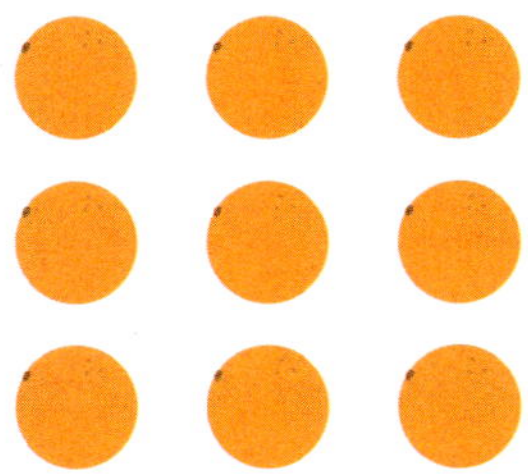

___ - ___ = ___

Kangaroo Contest

Help each kangaroo jump through the grass. Starting at the larger number in each sentence, draw the jumps for the number you need to subtract. Circle the number you land on and write it to finish the number sentence.

8 - 3 = ___

10 - 6 = ___

4 - 1 = ___

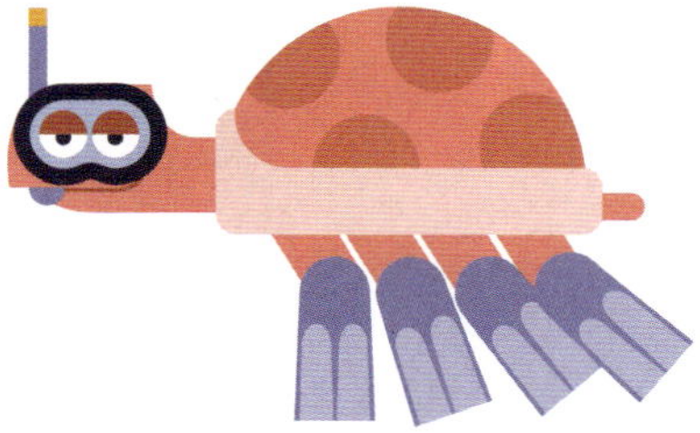

Find the Difference

Look at the pictures in each box. Use vertical subtraction to show the difference.

$$\begin{array}{r} 2 \\ -\,1 \\ \hline \end{array}$$

$$\begin{array}{r} 4 \\ -\,2 \\ \hline \end{array}$$

$$\begin{array}{r} 5 \\ -\,3 \\ \hline \end{array}$$

$$\begin{array}{r} 6 \\ -\,2 \\ \hline \end{array}$$

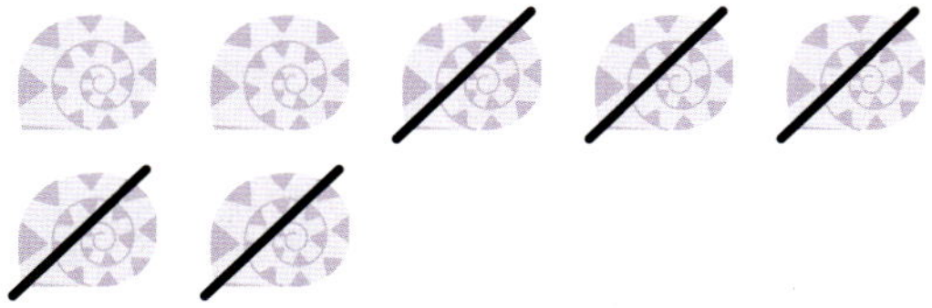

$$\begin{array}{r} 7 \\ -\,5 \\ \hline \end{array}$$

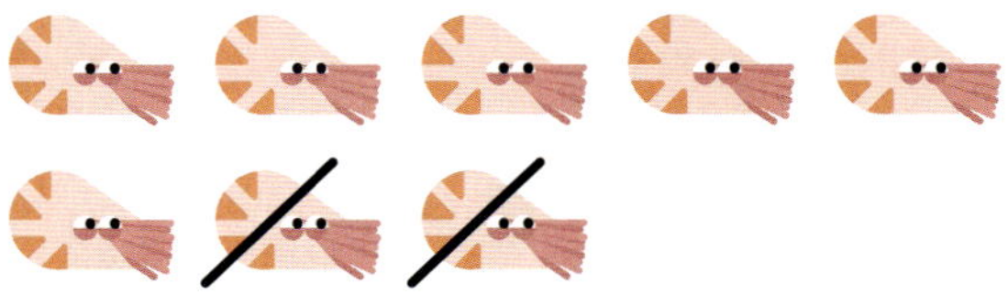

$$\begin{array}{r} 8 \\ -\,2 \\ \hline \end{array}$$

$$\begin{array}{r} 9 \\ -\,6 \\ \hline \end{array}$$

$$\begin{array}{r} 10 \\ -\,3 \\ \hline \end{array}$$

Celebrate finishing the review with a star sticker.

Add in Any Order

Flip It!

When adding, the order of the numbers does not matter—you can flip them, and the sum stays the same.

	4 + 2 = 6
	2 + 4 = 6

Play Ball

In each row, look at the number sentence in the first column. Color the ball with the same sum.

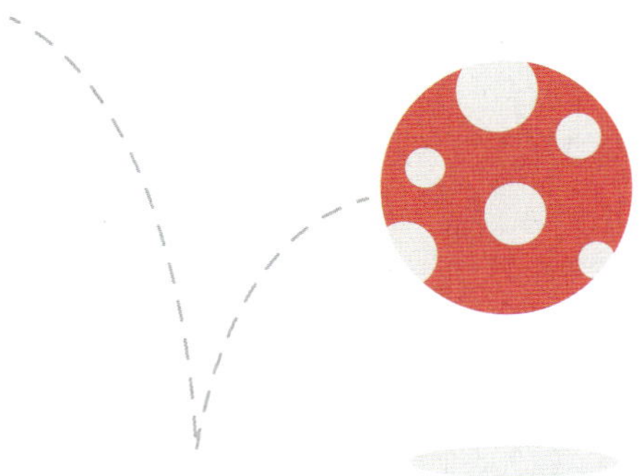

6 + 2	4 + 3	2 + 6	2 + 2
8 + 5	5 + 7	5 + 5	5 + 8
3 + 7	7 + 3	3 + 6	7 + 8

The *commutative property* of addition states that numbers can be added in any order without changing the sum.

Swap It!

For each set of pictures, write an addition sentence that matches.

 ___ + ___ = ___	 ___ + ___ = ___
 ___ + ___ = ___	 ___ + ___ = ___
 ___ + ___ = ___	 ___ + ___ = ___

Understanding that the order of numbers can be switched will be important when learning the upcoming *counting on strategy* for addition.

Launch  Introducing the Counting On Strategy

The *counting on strategy* is a faster way to add numbers together because you don't have to start counting from 1. A number line can help you!

Number Line

Start with the greater number and count on (or up) the number you want to add. For example, to find the sum of 6 + 3, start at 6 and jump forward 3 spaces to stop at the number 9.

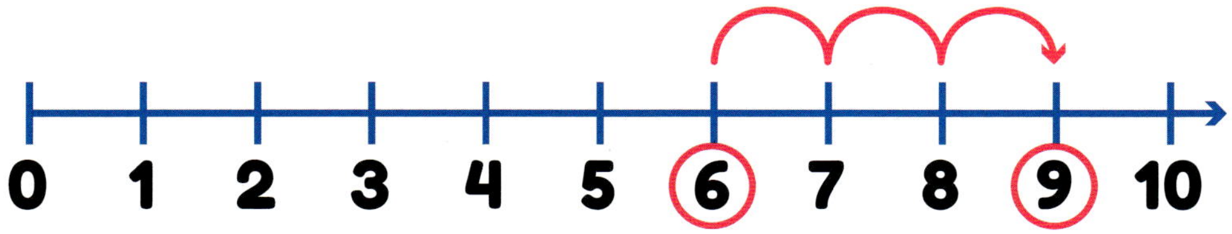

6 + 3 = 9

Ten Frame

You can also count on using a ten frame, which shows numbers in groups of 10.

Start with the greater number. Since this is a filled ten frame, you can start counting at 10. Count on (or up) the number you want to add. For example, to find the sum of 10 + 4, start at 10 and count 4 more to reach 14.

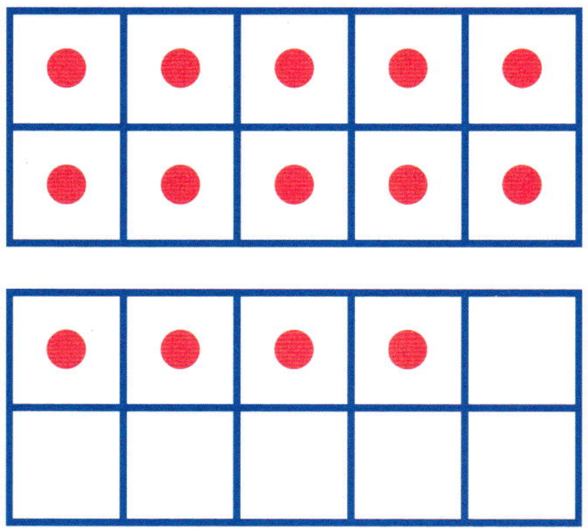

10 + 4 = 14

Counting on (or up) is a beginning mental-math strategy. Previously, your child likely counted **all** of the numbers in any two groups being added together. This strategy is a more efficient way to add.

Green Eggs

Draw eggs to finish the ten frames so that they match the number sentences. Then write the sum. Remember to start counting at 10.

10 + 3 = ___

10 + 2 = ___

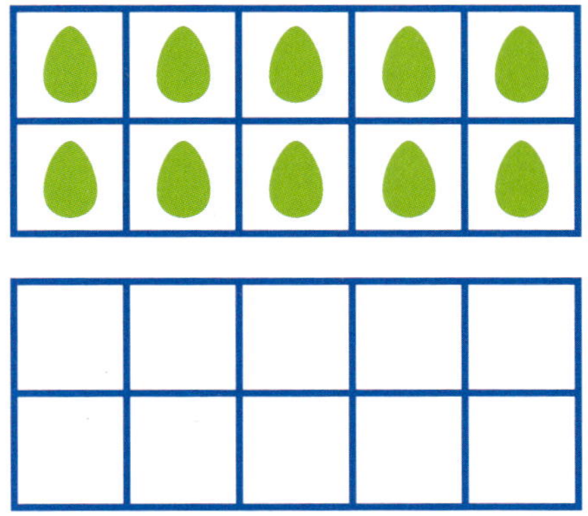

10 + 1 = ___

10 + 8 = ___

10 + 5 = ___

10 + 7 = ___

Materials needed: green crayon, pencil, or marker

Doubles + 1

Doubles Facts

In math, *doubles facts* are number sentences that show the same number added to itself. Notice how the first two numbers are repeated.

1 + 1 = 2	2 + 2 = 4	3 + 3 = 6	4 + 4 = 8	5 + 5 = 10
6 + 6 = 12	7 + 7 = 14	8 + 8 = 16	9 + 9 = 18	10 + 10 = 20

Doubles + 1 Strategy

When you have a number sentence that is almost a double, you can use the *doubles + 1* strategy.

Notice the number 6 is 1 more than 5.	5 + 6
Write the sentence as a doubles fact + 1.	5 + 5 + 1
Add the doubles together.	10 + 1
The sum is 1 more than the doubles fact.	11

Many children use their fingers or physical counters when learning to add. Knowing doubles facts and using strategies such as doubles +/- 1 can help your child make the shift to mental math.

Draw the Dots

- Count the dots on the first bug in the *Doubles* column and draw the same number of dots on the second bug. Then write how many dots there are in all.
- Do the same in the *Doubles + 1* column, but draw one extra dot on the second bug.
- Write how many dots in all. The first one has been done for you.

Doubles	How many dots?
	6

Doubles + 1	How many dots?
	7

Launch

Counting On Using a Number Line

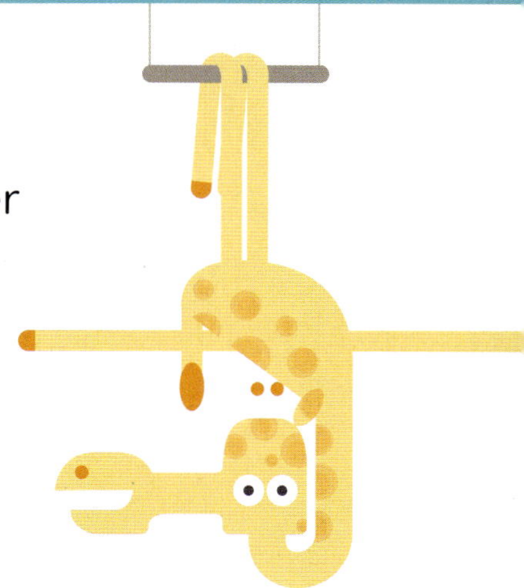

Flip a Switch

When using a number line to solve addition problems, remember to switch the order of the numbers being added if needed. That way, you can start counting on from the larger number.

4 + 14 = ?

14 + 4 = ?

Who's First?

Look at the number sentences below. Circle the number you would start with in order to count on.

3 + 15	7 + 11	6 + 14
2 + 5	10 + 8	18 + 2

Riddle: What season does the plus sign like the best?

Answer: SUM-mer!

Learning different ways to add and subtract allows your child to develop strategies for efficient problem solving. This helps to build true math fluency.

Wild Waves

Help each dolphin ride the waves. Look at each addition sentence. Starting at the larger number in the sentence, draw a jump to the next wave. Keep jumping forward the number you need to add. Circle the last wave and write the answer to finish the sentence.

17 + 3 = ___

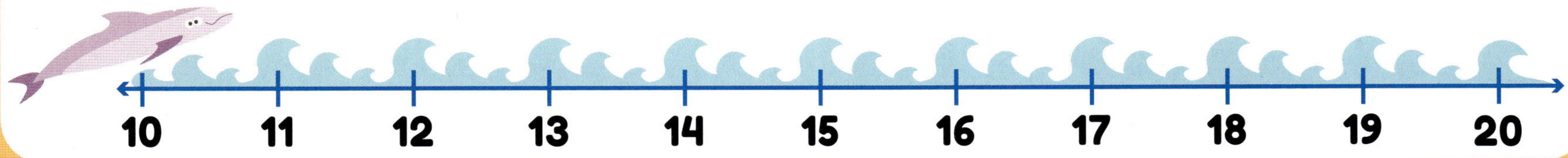

5 + 13 = ___

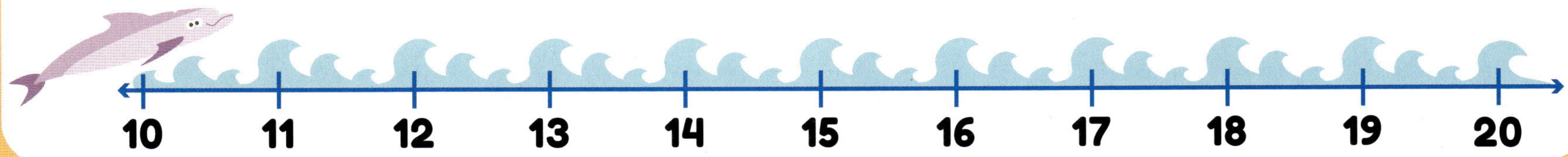

2 + 12 = ___

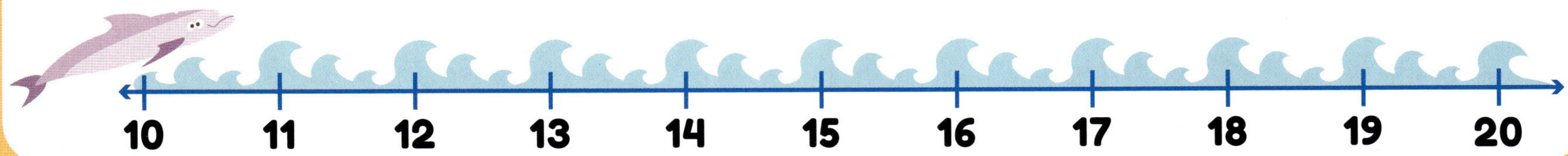

Counting On Strategy Practice

Nice Dice

Use the counting on strategy to find the sum for each number sentence. Circle the starting number on the number line, then draw the number of jumps needed to reach the sum. The first one has been done for you.

It is faster to start at the larger number and count on because there are fewer to count.

7 +		
2 +	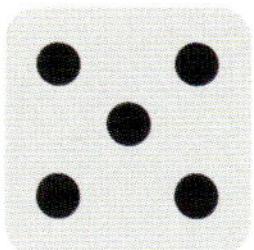	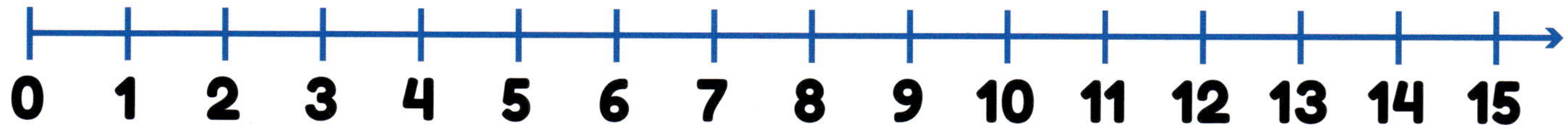
3 +		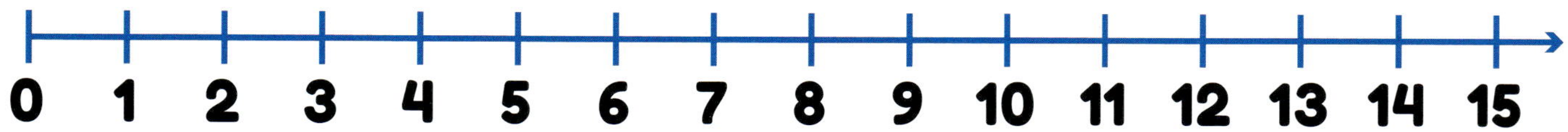
11 +		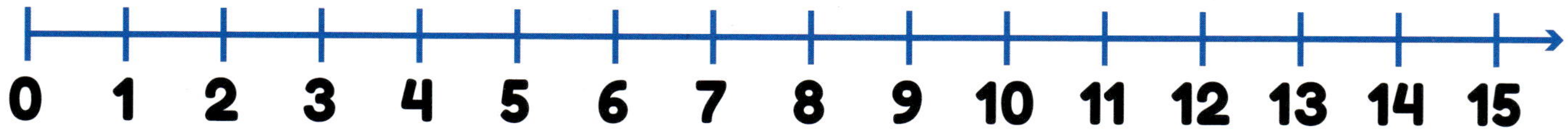

Tool Time

Number lines and ten frames are useful tools for addition. In each row, circle the number sentence that matches the tool next to it. Cross out the sentence that does not match. The first one has been done for you.

14 + 3 = 17 ~~12 + 3 = 15~~	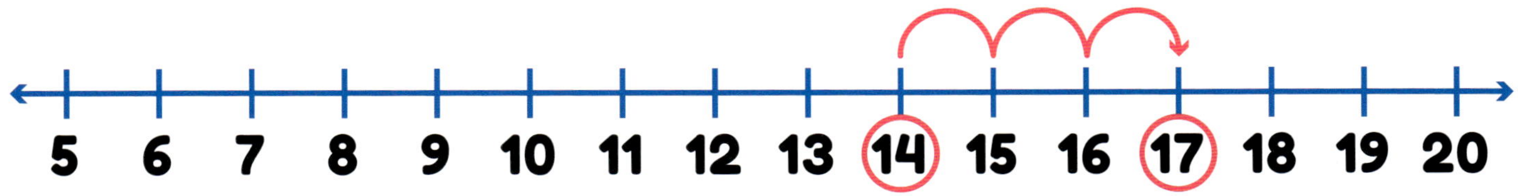
8 + 7 = 15 8 + 5 = 13	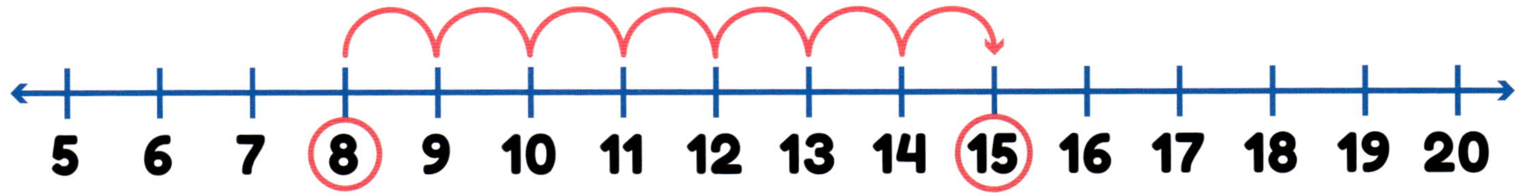
10 + 6 = 16 13 + 2 = 15	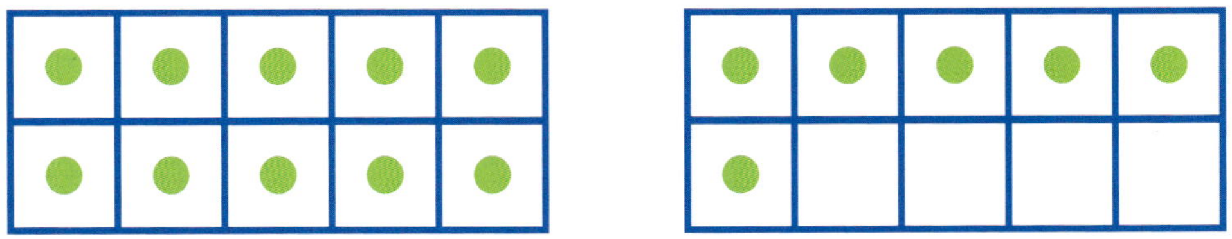
6 + 8 = 14 10 + 3 = 13	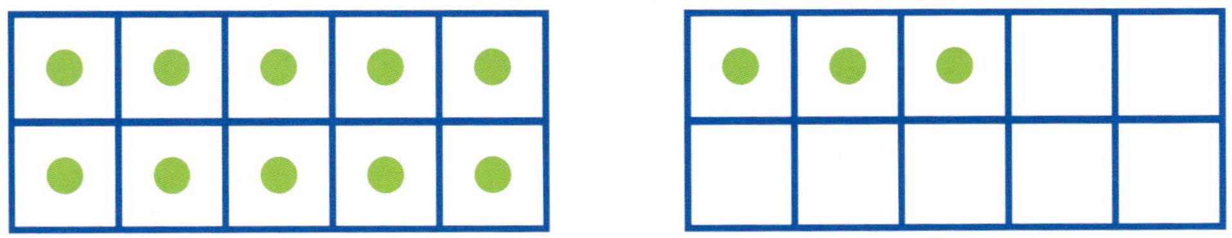

Text Me

Read the text messages below. To help answer the questions, use either a ten frame or number line on the next page.

How many points did Emma and Olivia make all together?

☐

How many people are invited to the party in all?

☐

How many eggs are needed all together?

☐

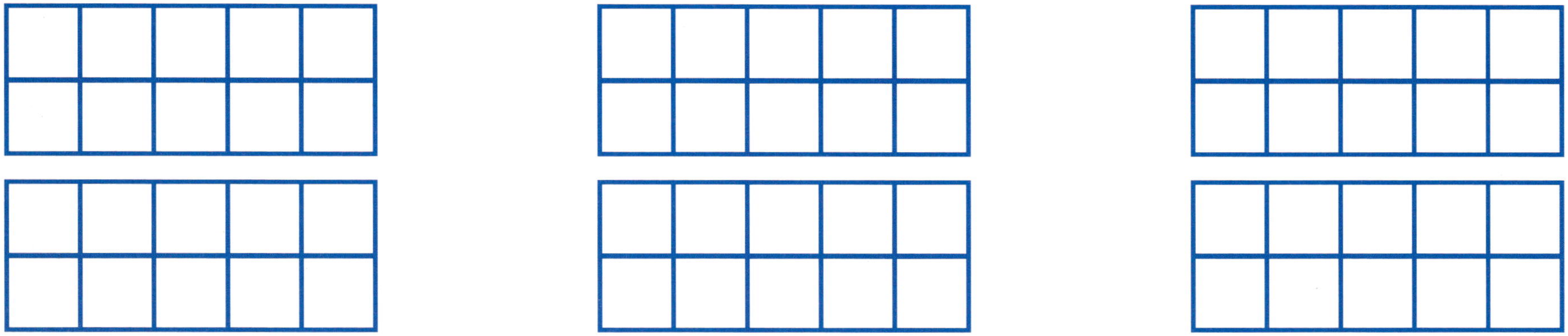

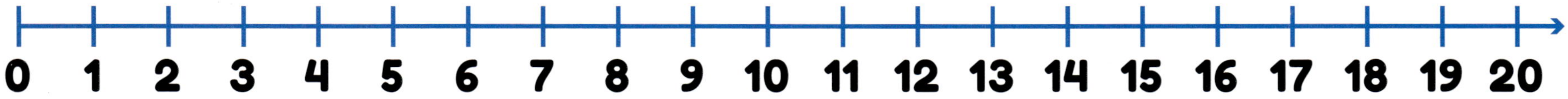
0 1 2 3 4 5 6 7 8 9 10 11 12 13 14 15 16 17 18 19 20

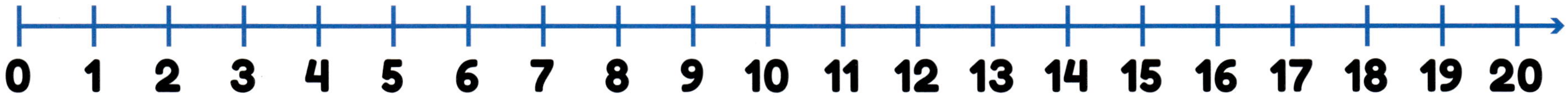
0 1 2 3 4 5 6 7 8 9 10 11 12 13 14 15 16 17 18 19 20

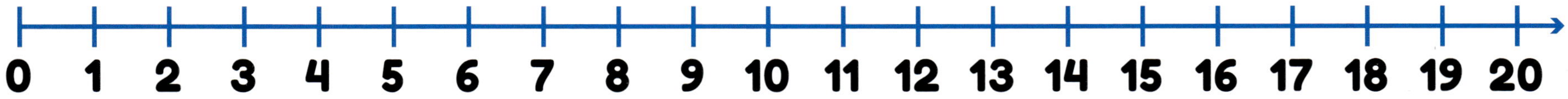
0 1 2 3 4 5 6 7 8 9 10 11 12 13 14 15 16 17 18 19 20

Game 1

12 + 6 = ___	7 + 4 = ___
9 + 6 = ___	3 + 11 = ___
8 + 5 = ___	1 + 9 = ___

Bingo Round 1

Find the sum for each problem. Use the number line to help you. Then color in each sum on the bingo cards. Any X is a free space. Which card has five colored spaces in a row to win at bingo?

B	I	N	G	O
11	X	10	14	18
1	7	9	16	X
4	6	X	15	17
2	8	3	X	19
X	5	12	13	20

B	I	N	G	O
4	8	12	16	X
X	6	10	13	18
2	7	X	15	17
1	5	9	X	20
3	X	11	14	19

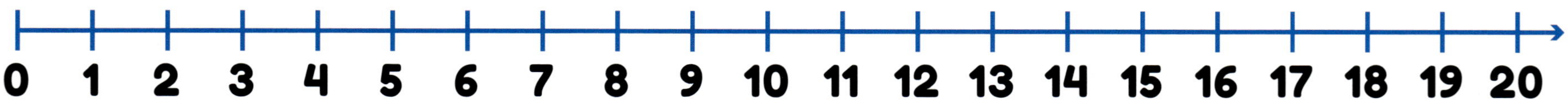

Game 2

2 + 1 = ___	9 + 10 = ___
2 + 2 = ___	3 + 3 = ___
7 + 5 = ___	8 + 12 = ___
14 + 4 = ___	2 + 3 = ___

Bingo Round 2

Play again! Find the sum for each problem. Use the number line to help you. Color in each sum on the bingo cards. Any X is a free space. Which card has colored spaces that form an X to win at bingo?

B	I	N	G	O
3	X	10	14	17
1	7	9	16	X
4	6	X	15	18
2	8	11	X	19
X	5	12	13	20

B	I	N	G	O
4	8	13	16	X
X	6	10	12	17
2	7	X	15	18
1	5	9	X	20
3	X	11	14	19

0 1 2 3 4 5 6 7 8 9 10 11 12 13 14 15 16 17 18 19 20

Read

The Carnival

Your child is ready to read *The Carnival*, a math-based storybook that features different strategies for solving addition problems.

During reading:

Ask your child to look at the number lines on pages 10 and 11. They should be able to easily recognize that one number line is much faster to count than the other. You can remind them that starting with the larger of two numbers requires fewer "jumps" along the number line.

After reading:

Ask your child which carnival ride they like best and why.

Look online for photos of carnival rides such as a Ferris wheel or roller coaster. Ask your child to count the number of cars they see on each ride.

Put together a jar of jelly beans (like the one in the story) or other small objects. Ask your child if they can guess how many are in the jar. After their guess, work together to count out separate groups of 10. Arrange them as if they were in a ten frame. How many are there in all? Was your child's guess close?

Celebrate your child's progress by placing a star sticker in the workbook.

Music Time

Find the sum for each problem in the Color Key. Then put the numbers in order from least to greatest in the boxes on the xylophone. Use the Color Key to color in the xylophone.

Color Key

10 + 6 = ___	**yellow**	**10 + 8 =** ___	**green**
10 + 3 = ___	**red**	**10 + 10 =** ___	**purple**
10 + 9 = ___	**blue**	**10 + 5 =** ___	**orange**

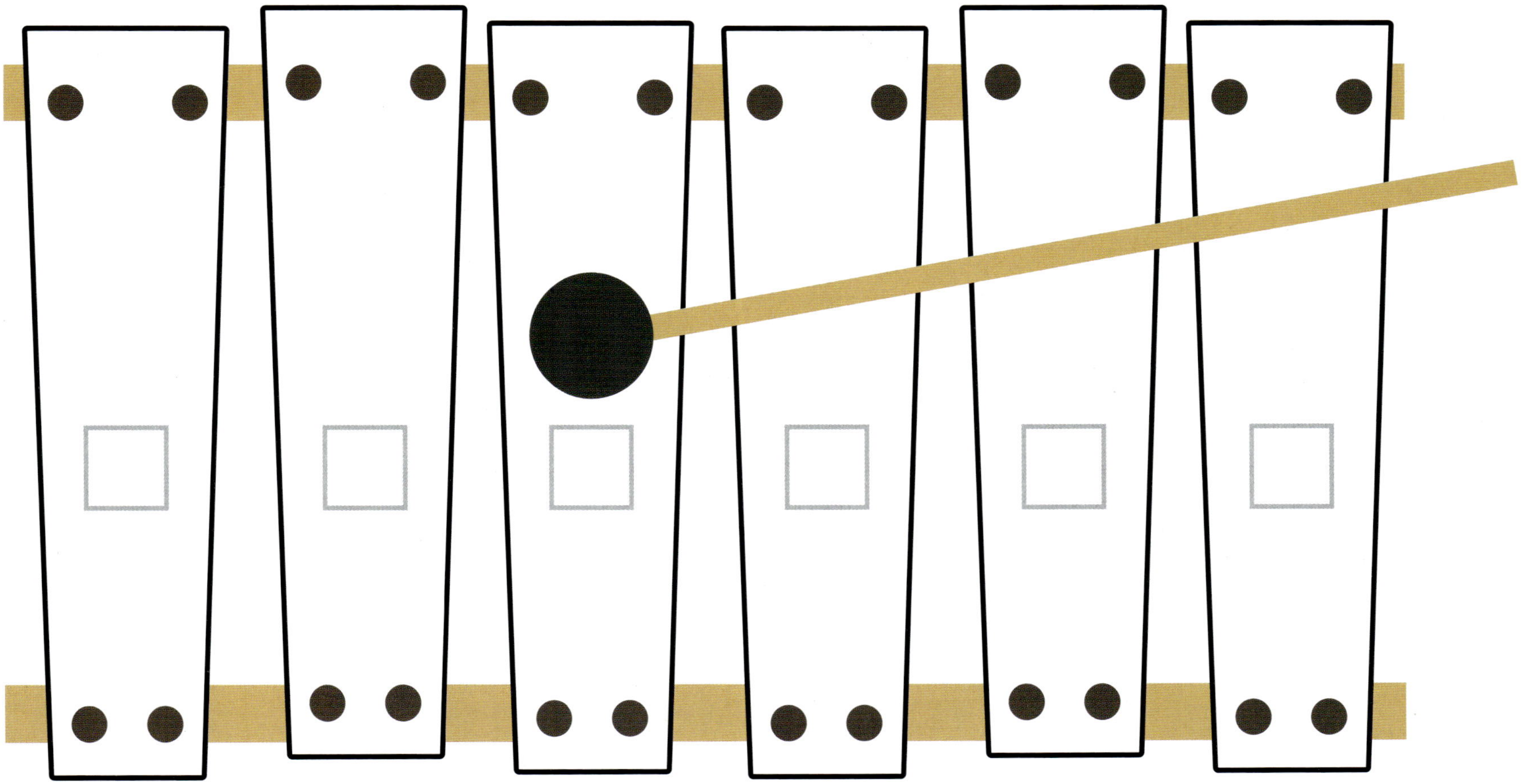

Your child has completed Step 6!
Celebrate by placing a star sticker in the workbook.

Counting Back Check-up

With the *counting back strategy*, you can use a number line to count back from a larger number to a smaller number. The number line can also be used to check your answer. Just flip the last two numbers!

12 - 5 = 7 Start at the larger number and count back 5. The answer is 7.	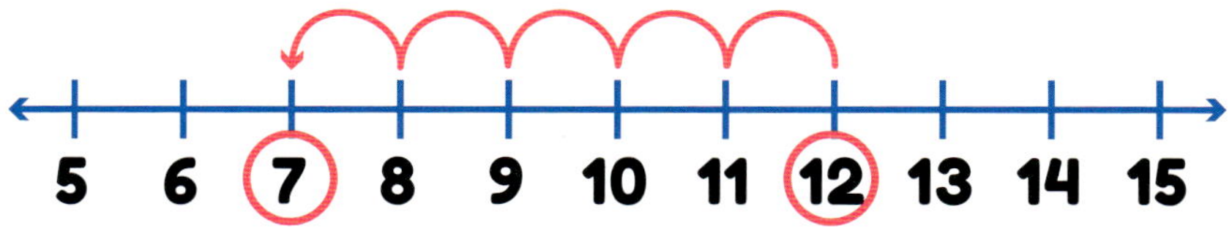
12 - 7 = 5 To check your answer, start at the larger number again. This time, count back by the **answer**. If you land on the original counting back number, your answer is correct.	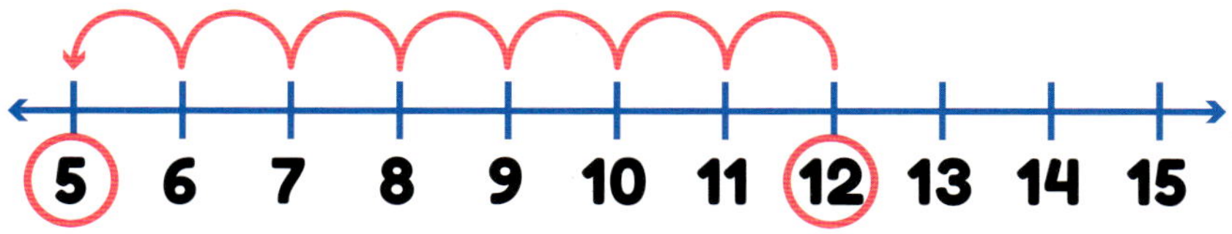

Flip Flop

Look at the subtraction sentence in each box. How would you flip it to check your answer? Circle it!

19 - 13 = 6	**6 - 13 = 19**	**19 - 6 = 13**
15 - 12 = 3	**15 - 3 = 12**	**12 - 15 = 3**

Back Track

Draw the first subtraction sentence on the number line and write the answer. Check your answer using the second number line. Write the flipped sentence.

12 - 3 = ___	3 4 5 6 7 8 9 10 11 12 13 14 15
___ - ___ = ___	3 4 5 6 7 8 9 10 11 12 13 14 15
14 - 8 = ___	3 4 5 6 7 8 9 10 11 12 13 14 15
___ - ___ = ___	3 4 5 6 7 8 9 10 11 12 13 14 15

Part-Part-Whole Strategy

The *part-part-whole strategy* can be used to solve either addition or subtraction problems.

The whole is the entire group of something. 	The parts are the smaller pieces that make up the whole. 

Remember that each part is always smaller than the whole. The parts do not have to be the same size or even the same thing.

WHOLE

part

Whole 10	
Part **3**	**Part** **7**

3 + 7 = 10 For addition, you take all the parts and put them together to make the whole group.	**part + part = WHOLE**
10 - 3 = 7 For subtraction, you take the whole group and break it down into its smaller parts.	**WHOLE - part = part**

Mastery of this concept will help your child understand word problems in which they must find the unknown, which might be the whole or one of the parts.

Puzzling Parts

Find and circle the puzzle piece that completes each puzzle. Use the number line to help you. Then write the subtraction sentence for each puzzle.

Whole 12 | Part 8 | Part ?

___ - ___ = ___

Part 4 | Part 6 | Part 8

Whole 17 | Part ? | Part 5

___ - ___ = ___

Part 11 | Part 12 | Part 7

Whole 19 | Part 4 | Part ?

___ - ___ = ___

Part 17 | Part 10 | Part 15

0 1 2 3 4 5 6 7 8 9 10 11 12 13 14 15 16 17 18 19 20

Launch

Missing Doubles

Doubles facts for subtraction are the opposite of the doubles facts for addition. When subtracting, just flip the sentence around!

Addition	Subtraction
1 + 1 = 2	2 - 1 = 1
2 + 2 = 4	4 - 2 = 2
3 + 3 = 6	6 - 3 = 3

Addition	Subtraction
4 + 4 = 8	8 - 4 = 4
5 + 5 = 10	10 - 5 = 5
6 + 6 = 12	12 - 6 = 6

Addition	Subtraction
7 + 7 = 14	14 - 7 = 7
8 + 8 = 16	16 - 8 = 8
9 + 9 = 18	18 - 9 = 9

Splitting Pairs

If you know the sum of a doubles fact, you can use the part-part-whole strategy to find the missing numbers.

The parts for a doubles fact are two of the same number that equal the whole. Draw a picture of the whole.

Whole 8	
Part ?	Part ?

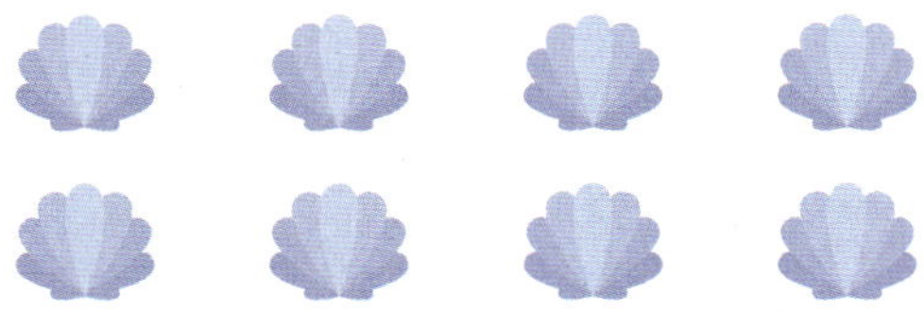

Split the shells into 2 equal groups. Both groups should have 4 shells each.

Whole 8	
Part 4	Part 4

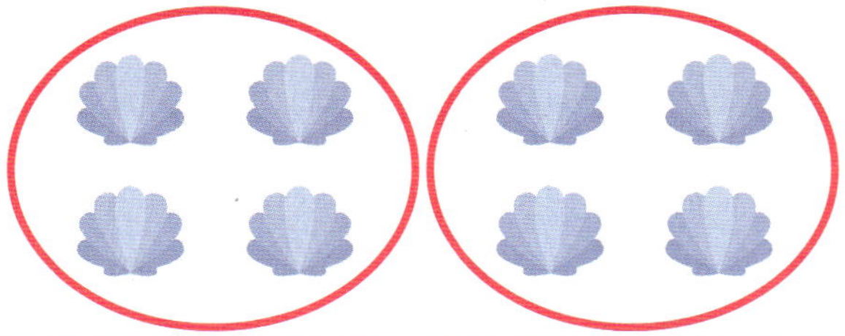

Mastering doubles facts will help your child with skip counting, repeated addition, and arrays—as well as multiplication and division.

Dinner for Two

Kazoo is making a salad for Batu. They will both eat the same amount. Circle two equal groups to find the parts. Write the missing numbers in the part-part-whole box. The first one has been done for you.

There are 8 celery stalks.
How many will each pal eat?

Whole 8	
Part 4	Part 4

There are 10 carrots.
How many will each pal eat?

Whole 10	
Part	Part

There are 12 olives.
How many will each pal eat?

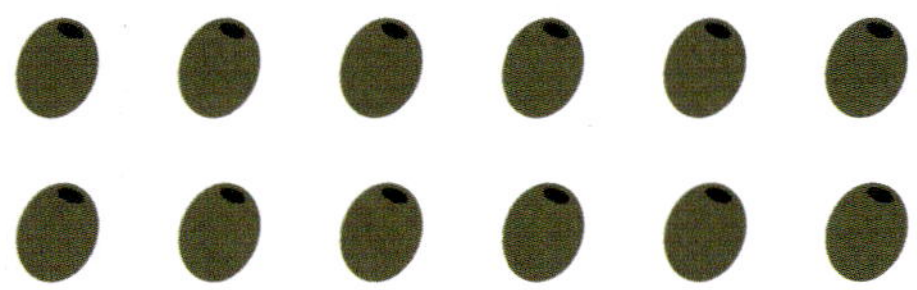

Whole 12	
Part	Part

There are 6 onions.
How many will each pal eat?

Whole 6	
Part	Part

Finding Missing Numbers

Think It Through Mystery

Be a number detective. What can you do if you know the answer but need to find the missing numbers? If you have some clues, you can find them!

Mystery Sentence
? - ? = 12

Clues: 6, 18, 20

Step 1: Remember that the first number in the sentence must be the same or more than the answer, or difference. Try each clue as the **first** number.

20 - ? = 12
18 - ? = 12
~~6 - ? = 12~~

Step 2: Look at the possible sentences from Step 1. You know the missing number **must be the same or less than** the first number. Try the remaining clues.

20 - 18 = 12
20 - 6 = 12
~~18 - 20 = 12~~
18 - 6 = 12

Step 3: Look at each sentence. Cross out any that are not true. You can use a number line to check each one.

~~20 - 18 = 12~~
~~20 - 6 = 12~~
18 - 6 = 12

You found the answer!

When your child needs to find the missing number instead of a difference, it can help to remember that a subtraction sentence always starts with the largest number.

Plug It In!

Find the missing numbers in the number sentence. Use the clues and the number line to help you. The first one has been done for you.

Mystery Sentence	Clues	Try It!	True!
? - ? = 12	19, 7, 15	~~19 - 15 = 12~~ 19 - 7 = 12 ~~15 - 7 = 12~~	19 - 7 = 12
? - ? = 14	11, 3, 17		
? - ? = 13	13, 0, 6		
? - ? = 4	5, 7, 11		

This lesson focuses on problem-solving and mathematical reasoning skills. It encourages your child to develop a deeper understanding of the processes used to arrive at solutions, rather than remembering and applying a set of procedures.

0 1 2 3 4 5 6 7 8 9 10 11 12 13 14 15 16 17 18 19 20

Treasure Hunt

Find the missing part in each box. Use the number line to help you. Write the answer in the box and in the sentence. Then, starting at the arrow, follow the clues on this page to find the treasure on the next page.

Whole 13	
Part 8	Part

Clue 1:
Move ____ spaces right.

Whole 18	
Part	Part 16

Clue 2:
Move ____ spaces down.

Whole 15	
Part 11	Part

Clue 3:
Move ____ spaces left.

Whole 19	
Part 13	Part

Clue 4:
Move ____ spaces down.

Whole 14	
Part	Part 9

Clue 5:
Move ____ spaces right.

0 1 2 3 4 5 6 7 8 9 10 11 12 13 14 15 16 17 18 19 20

Which Way?

To reach the treasure, follow the clues from the previous page. Watch out for sea creatures!

Read — A Day at the Beach

Your child is ready to read the math-based activity book *A Day at the Beach*. Use the sticker sheet marked "A Day at the Beach" that came with this pack.

During reading:

Remind your child of the basic concept behind the part-part-whole strategy: "parts" are the pieces of something, while the "whole" is the entire thing. Parts can never be larger than the whole. This strategy can be used when solving addition or subtraction problems.

It may also be helpful to remind your child of the fact families they learned about in earlier steps, since part-part-whole is a continuation of the same idea.

After reading:

Does your child have a favorite number? Ask them to create a number bond chart like the ones in the book, using their favorite number. How many ways can they represent this number?

What is your child's favorite flavor of ice cream? Can they imagine an ice cream cone with 20 scoops? Ask your child to draw it.

Celebrate your child's progress by placing a star sticker in the workbook.

Math about Me!

My first name is ______________________. It has ☐ letters.

My last name is ______________________. It has ☐ letters.

Write an addition sentence using these numbers.
You can use the number line to help you add them up. ☐ + ☐ = ☐

I am ☐ years old. I have lost ☐ teeth.

Write an addition sentence using these numbers.
Use the number line to help you add them up. ☐ + ☐ = ☐

There are ☐ people living in my home, including me. I have ☐ pets.

Write an addition sentence using these numbers.
Use the number line to help you add them up. ☐ + ☐ = ☐

0 1 2 3 4 5 6 7 8 9 10 11 12 13 14 15 16 17 18 19 20

Your child has completed Step 7!
Celebrate by placing a star sticker in the workbook.

HOOKED ON MATH

It's Time to Celebrate!

____

has completed Math Fundamentals Steps 6 & 7
and is now ready to move on to the
Math Fundamentals Steps 8 & 9 Workbook!

6

9

SKILLS MASTERED

- Add in any order
- Counting on
- Doubles + 1
- Counting back
- Part-part-whole
- Missing doubles
- Finding missing numbers

7

8

Don't forget to share your success and tag us on social media so we can celebrate with you!

Credits

Design Team
Julie Linden Frazier
Lindsey Frazier

Editorial Team
Donna Giachetti
Cynthia Stierle
Austin Feld

Illustrations
Dubit Limited

Educational Advisor
Lesley Freed, Curriculum Specialist

Published by Sandviks HOP, Inc.

Printed in China

HOOKED ON MATH

Math Fundamentals

18025 (05/22)

HOOKED ON MATH

Math Fundamentals

Steps 8 & 9

Workbook

Tips for Success

- Math Fundamentals introduces your child to the building blocks of math—number sense, addition, and subtraction.
- Work at a pace that's right for you and your child.
- Fit your session length to your child's attention span. It's okay to work with the App one day and this workbook the next.
- Before beginning, flip through the workbook and note any materials you might need. You should plan to have pencils, a pair of dice, and crayons ready.
- You'll see parent tips throughout this workbook. These tips are designed to answer questions you or your child might have about the learning.
- Repetition and review are essential to basic math fluency. Each workbook step builds on the ones before, so make sure your child has mastered the current lesson before moving forward.
- As you go through the workbook, encourage your child to explore the connection between numbers and the real world. Reinforce concepts with familiar household objects such as toys or snacks to make adding and subtracting more memorable.
- Have fun with it!

HOOKED ON MATH

Math Fundamentals

Steps 8 & 9

Workbook

This workbook and the included storybooks introduce fundamental math skills.

Step 8: Numbers to 120

- Introducing the hundred chart
- Ordering three-digit numbers
- Place value: tens and ones

Step 9: Skip Counting

- Skip counting by 2s on a number line
- Skip counting by 2s on a hundred chart
- Skip counting by 5s
- Skip counting by 10s

Contents

Lesson Overview

Welcome! This workbook provides supplemental activities that will solidify and expand the learning your child is enjoying with the Hooked on Math App. Follow the easy, step-by-step instructions to guide your child's exploration of numbers and operations. Every lesson helps your child build a strong foundation in fundamental math skills while having fun!

1. Launch

Your child will be introduced to math concepts and skills through interactive lessons in the App on your mobile device or Windows or Mac computer. Download the App at:
hookedonmath.com/get-the-app

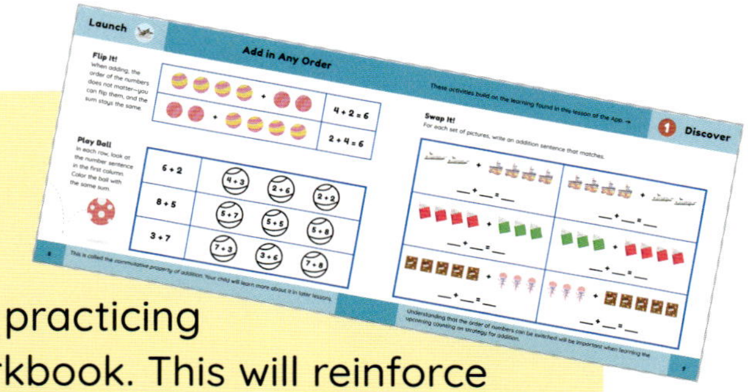

2. Discover

Next, your child will have fun exploring math concepts and practicing new skills in the workbook. This will reinforce the App lessons and provide a tangible application of the skills learned.

3. Play

In addition to the workbook activities, your child will enjoy two math-based storybooks that will strengthen their learning and add to the fun!

4. Celebrate

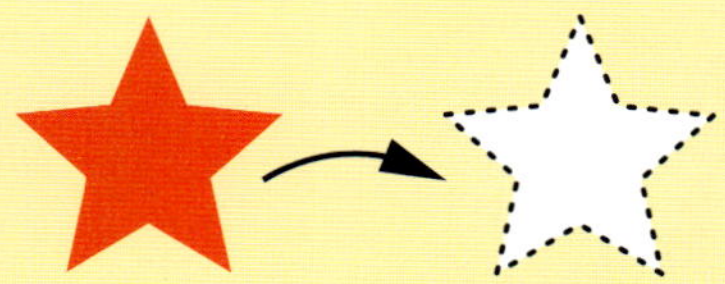

Celebrate your child's progress by placing a sticker in the workbook as learning milestones are reached. This also provides a helpful marker to remember where to begin next time.

Special notes are included in this space to provide additional context related to what is being taught.

All Ashore!

- Pick a pal. Starting at 0, roll the die and count forward that number.
- Circle the number you land on. From there, keep rolling until you land on or pass 20.
- Write how many times you rolled by counting the circled numbers.
- Play again. Who reaches the beach in the fewest rolls?

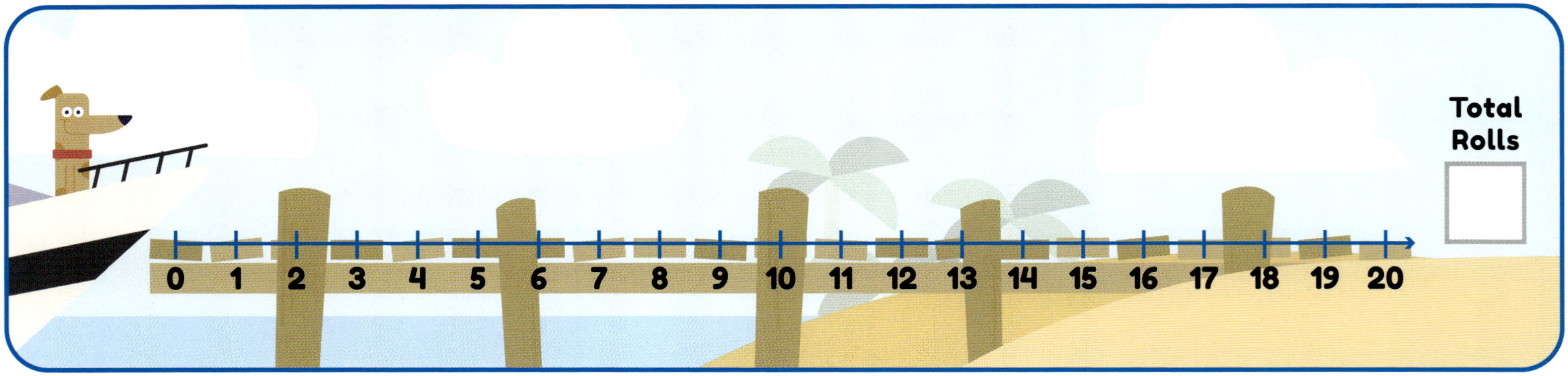

Materials needed: Dice

All Aboard!

- Pick a pal. Starting at 20, roll the die and count back that number.
- Circle the number you land on. From there, keep rolling until you land on or pass 0.
- Write how many times you rolled by counting the circled numbers.
- Play again. Who reaches the ship in the fewest rolls?

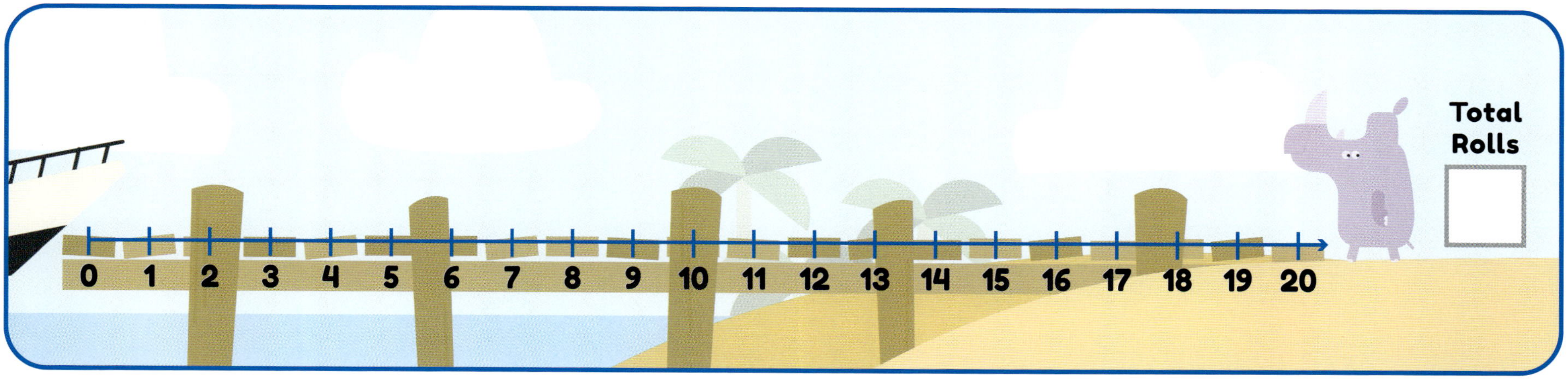

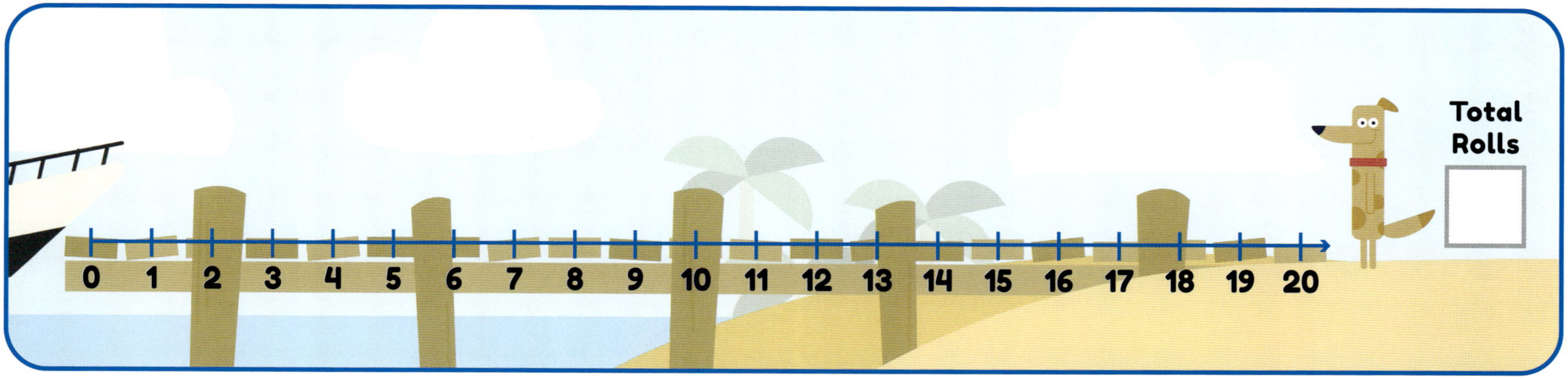

Double Bubble Trouble

Write the sum of the doubles fact in the bubbles above the tub. Then write the sum of the doubles + 1 fact on the tub. The first one has been done for you.

5 + 5 = 10

5 + 6 = 11

6 + 6 = ☐

6 + 7 = ☐

7 + 7 = ☐

7 + 8 = ☐

8 + 8 = ☐

8 + 9 = ☐

9 + 9 = ☐

9 + 10 = ☐

10 + 10 = ☐

10 + 11 = ☐

Special Delivery

Which box goes on the truck? Circle the box with the number that completes the part-part-whole number bond on each truck.

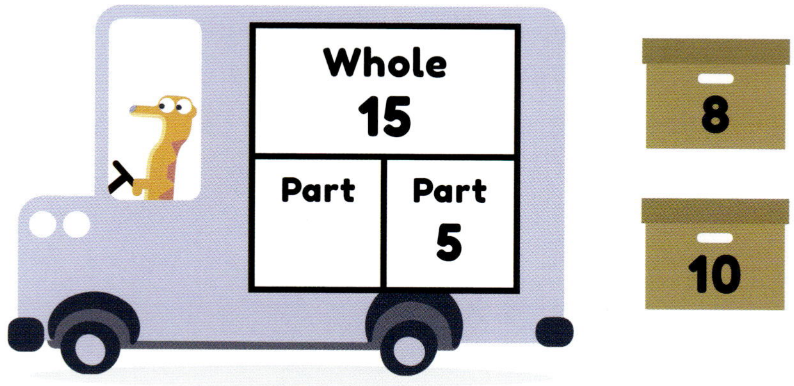

Addition Facts to 20 Review

Show What You Know!

For each row, read the fact in the **I KNOW** column, read the directions in the **I CAN** column, then write the answer in the **I DID IT!** column. The first one has been done for you.

I KNOW	I CAN	I DID IT!
How to write an addition sentence that shows two or more numbers added together to equal the sum.	Write an addition sentence, underline the plus sign, and circle the sum.	12 + 8 = 20
Numbers can be added in any order.	Flip 17 + 3 = 20 and write the number sentence.	
Addition problems can use pictures or numerals.	Write the number sentence that shows: 	
Number lines and ten frames are tools that help me add.	Write the number sentence that shows: 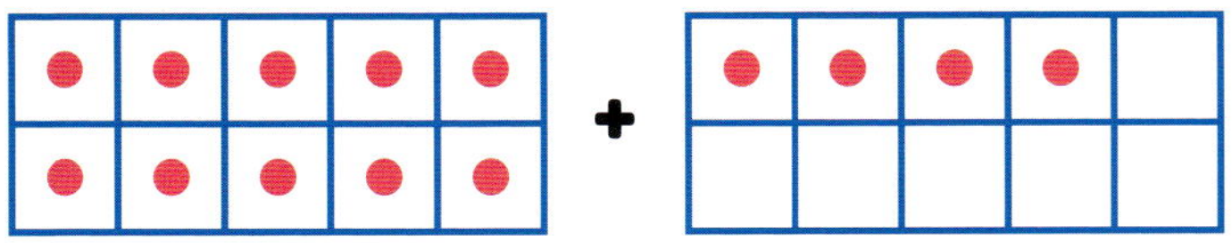	

Reviewing and practicing basic math concepts helps your child build connections that are the key to being able to automatically recall the information later.

Check the Facts

How many addition facts do you know? Write the sum for the facts in each box. Put a check mark at the end of the row if you know them all. Circle any facts you need to practice.

						Check
2 + 3	5 + 5	8 + 5	4 + 4	6 + 9	9 + 9	
6 + 4	2 + 7	10 + 0	4 + 5	7 + 12	6 + 11	
8 + 7	14 + 2	5 + 2	16 + 3	7 + 7	6 + 6	
1 + 7	10 + 3	8 + 8	2 + 8	11 + 4	18 + 1	

For an extra challenge for a child who has mastered the material, set a timer to see how quickly they can complete the problems. But remind your child that accuracy is more important than speed.

Launch

Introducing the Hundred Chart

Numbers don't stop at 20. A helpful math tool for working with larger numbers is called a *hundred chart*. It is a grid that shows the numbers from 1 to 100. It is like a number line organized in groups of 10.

Parts of the Chart

It starts with the number **1**. There are ten numbers in each row. Moving across from left to right, each number goes up by one: **1**, **2**, **3**, **4**, **5**, **6**, **7**, **8**, **9**, **10**.

+1 +1 +1 +1 +1 +1 +1 +1 +1

1	2	3	4	5	6	7	8	9	10
11	12	13	14	15	16	17	18	19	20
21	22	23	24	25	26	27	28	29	30
31	32	33	34	35	36	37	38	39	40
41	42	43	44	45	46	47	48	49	50
51	52	53	54	55	56	57	58	59	60
61	62	63	64	65	66	67	68	69	70
71	72	73	74	75	76	77	78	79	80
81	82	83	84	85	86	87	88	89	90
91	92	93	94	95	96	97	98	99	100

The second row starts with the number **11** and ends with **20**. Then the next row begins with **21** and ends with **30**.

The counting pattern repeats in every row, all the way to **100**.

Use the hundred chart to fill in the missing numbers.

43 ___ 45 ___ 47 48

71 ___ 73 ___ ___ 76

Challenge: Count to 100 by 1s. Point to and say each number aloud.

What's in the Sea?

Use the Color Key to color in the correct squares to reveal a hidden picture.

Color Key

Green: 70, 79, 80, 88, 89, 90

Brown: 91, 92, 93, 94, 95, 96, 97, 98, 99, 100

Pink: 5, 6, 14, 15, 16, 21, 22, 31, 32, 51, 52, 61, 62, 66, 76

Yellow: 25, 26, 27, 28, 34, 35, 36, 37, 38, 39, 42, 43, 44, 45, 46, 47, 49, 50, 54, 55, 56, 57, 58, 59, 65, 67, 68

Blue: all remaining squares

1	2	3	4	5	6	7	8	9	10
11	12	13	14	15	16	17	18	19	20
21	22	23	24	25	26	27	28	29	30
31	32	33	34	35	36	37	38	39	40
41	42	43	44	45	46	47	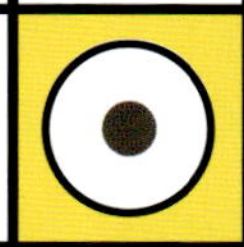	49	50
51	52	53	54	55	56	57	58	59	60
61	62	63	64	65	66	67	68	69	70
71	72	73	74	75	76	77	78	79	80
81	82	83	84	85	86	87	88	89	90
91	92	93	94	95	96	97	98	99	100

Materials needed: crayons or colored pencils in green, brown, pink, yellow, and blue

Hundred Chart Practice

A hundred chart is a useful tool for seeing number patterns.

A hundred chart makes it easy to add 10 to other numbers by going down the columns. Starting at **1**, you can count **1**, **11**, **21**, **31**, **41**, **51**, **61**, **71**, **81**, **91**. Notice that the first digit goes up by one, while the second digit remains the same.

1	2	3	4	5	6	7	8	9	10
11	12	13	14	15	16	17	18	19	20
21	22	23	24	25	26	27	28	29	30
31	32	33	34	35	36	37	38	39	40
41	42	43	44	45	46	47	48	49	50
51	52	53	54	55	56	57	58	59	60
61	62	63	64	65	66	67	68	69	70
71	72	73	74	75	76	77	78	79	80
81	82	83	84	85	86	87	88	89	90
91	92	93	94	95	96	97	98	99	100

In addition to ten numbers in each row, there are ten numbers in each column. Starting at **10**, you can count **10**, **20**, **30**, **40**, **50**, **60**, **70**, **80**, **90**, **100**.

Keep Going!

Numbers continue after 100. The counting sequence repeats itself, but you say *one hundred* before the rest of the number.

101	102	103	104	105	106	107	108	109	110
111	112	113	114	115	116	117	118	119	120
121	122	123	124	125	126	127	128	129	130

Although the focus here is mainly on one-digit and two-digit numbers, it is important for your child to understand that the same number patterns continue with three-digit numbers and beyond.

Chart Puzzle

Look at the sections of the hundred chart below. Add 1 to fill in the missing numbers going across. Add 10 to fill in the missing numbers going down. The first one has been done for you.

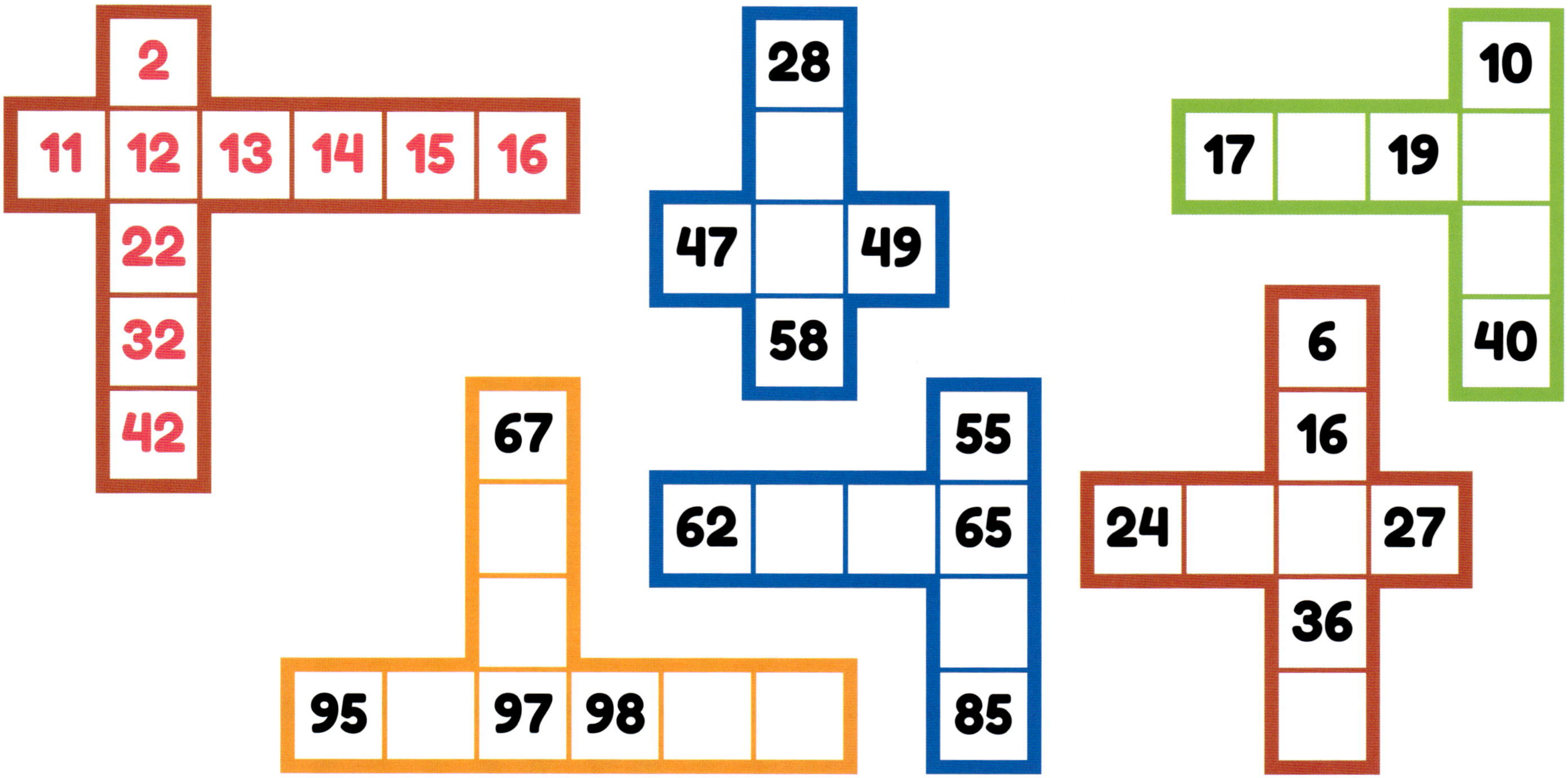

Ordering Three-digit Numbers

A *digit* is any number from zero to nine.

Numbers can have one digit, two digits, three digits, or more. The value of a digit is determined by its place, or position, in a number. The digit 3 can be found in 3, 30, and 300.

One Digit	3
Two Digits	30
Three Digits	300

Order Up!

Ordering numbers means to place them from lowest (or first) to highest (or last). When ordering numbers, you can compare the numbers based on how many digits they have.

- > greater than
- < less than
- = equal to, or the same

Circle the correct symbol to make each number sentence true.

5 > = < 24 **125 > = < 32** **99 > = < 99**

Compare-a-Spin

Spin the first spinner and write the number in the first column. Spin the second spinner and write the number in the second column. Then fill in the center circle with the symbol that makes the sentence true. The first one has been done for you.

$>$ $<$ $=$

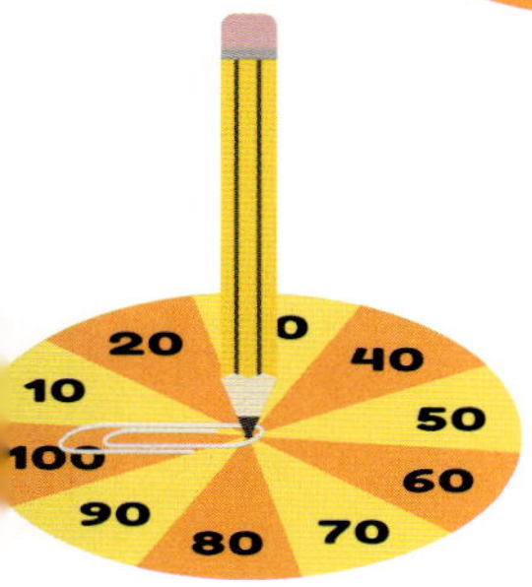

How to Make the Spinner: Place one end of the paper clip in the center of the circle. Put the pencil tip inside the paper clip on the center dot. Now flick the paper clip so it spins around the pencil.

30	$>$	10

Materials needed: sharpened pencil, paper clip

Launch 

Place Value: Tens and Ones

The value of a digit depends on where it is placed in the number.

In a two-digit number, there are two places: a *tens* place and a *ones* place. **There can only be one digit in each place.**

In the number **19**, there is 1 group of ten in the tens place, and 9 ones in the ones place.

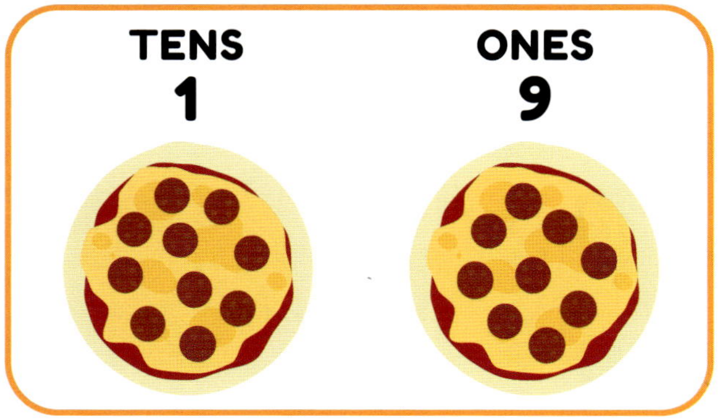

The digit in the tens place always represents a group of ten.

In the number **20**, there are 2 groups of ten in the tens place, and a 0 in the ones place because there are no ones.

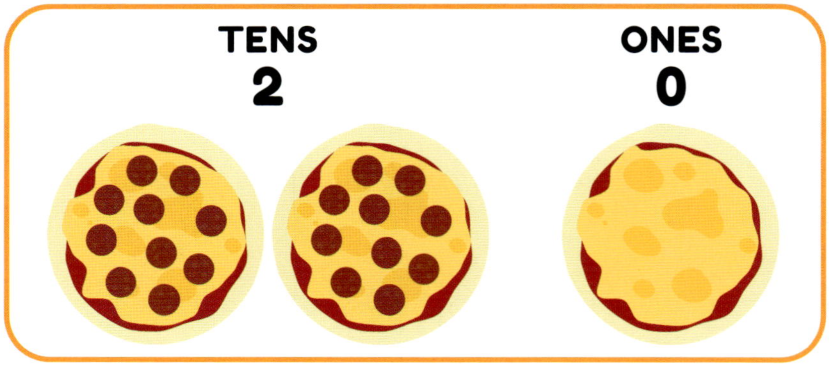

Top It!

Circle the number that matches the total number of toppings on the pizza.

TENS	ONES	Circle It
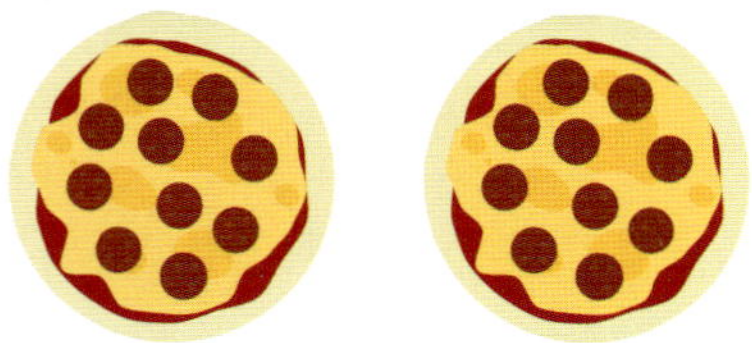		25 21 35

Remind your child that in Step 1 they learned that numbers such as 19 were "ten plus nine more."

Another way to look at two-digit numbers is with place value blocks.

		Tens	Ones
A single ones block equals the number 1.	 = 1		1
The number 9 is shown with 9 ones blocks.	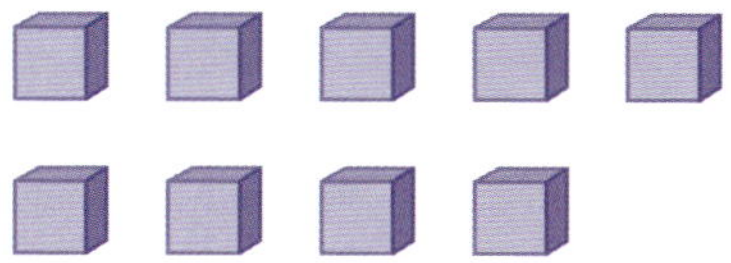 = 9		9
If one more block is added after 9 ones, you have a group of 10.	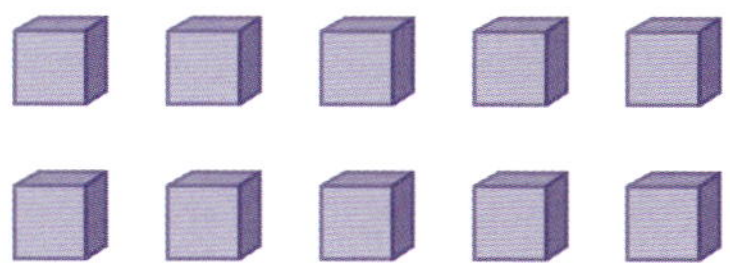 = 10	1	0
Ten ones blocks can be stacked together into a tens rod. Using a tens rod makes it easier to count groups of 10.	 = 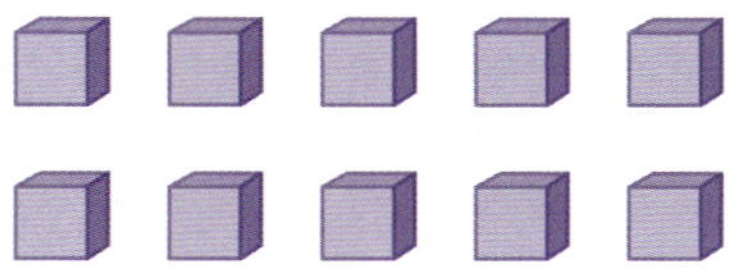 = 10	1	0

Discover

Count It!

In each row, circle the number that matches the number of tens rods and ones blocks.

TENS	ONES	Circle It
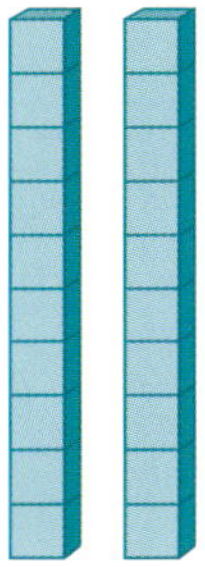	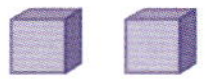	12 42 22
	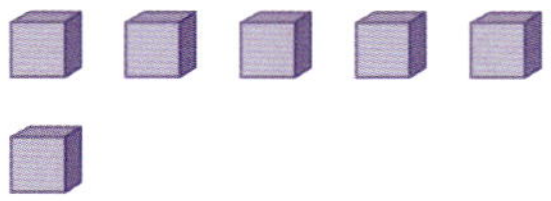	25 16 61
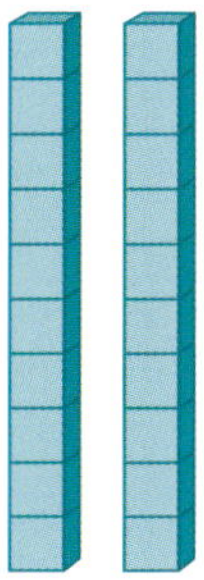	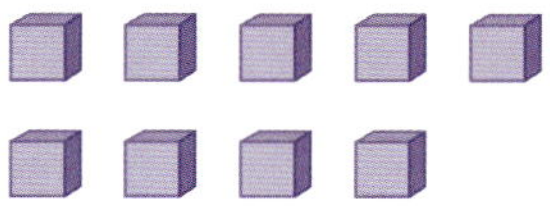	19 29 89

Color Match

Color the shells to match the place value blocks that show the same number.
The first one has been done for you.

Materials needed: crayons or colored pencils in green, yellow, red, orange, blue, and pink

Play

Color by Place Value

Color Key

Color the numbers with a 5 in the ones place	**blue**
Color the numbers with a 0 in the ones place	**green**
Color the numbers with a 3 in the tens place	**yellow**
Color the numbers with a 6 in the tens place	**orange**
Color all the three-digit numbers	**purple**

Color by Place Value

Color the numbers in the grid below according to the Color Key to reveal the hidden picture.

211	123	121	874	222	257	15	5	922
552	392	356	726	588	85	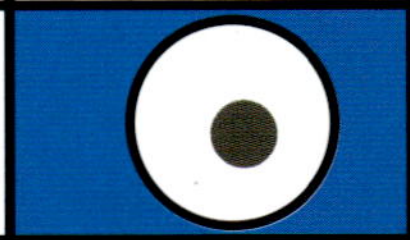	75	421
319	401	371	841	502	55	95	45	61
842	191	36	33	32	37	31	114	143
944	38	34	31	39	20	90	999	158
40	20	10	40	50	10	20	193	221
154	0	90	70	80	70	119	786	294
201	182	149	62	68	296	111	917	313

Materials needed: crayons or colored pencils in blue, green, yellow, orange, and purple

Question in Code

In each box, write the number of the place value blocks to finish the Code Key. Then use the key to fill in the answer to the riddle below.

Code Key

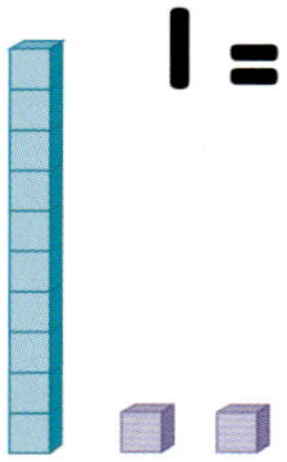 I = ___	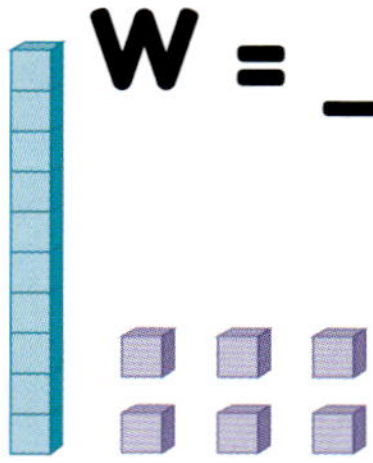 W = ___	 U = ___	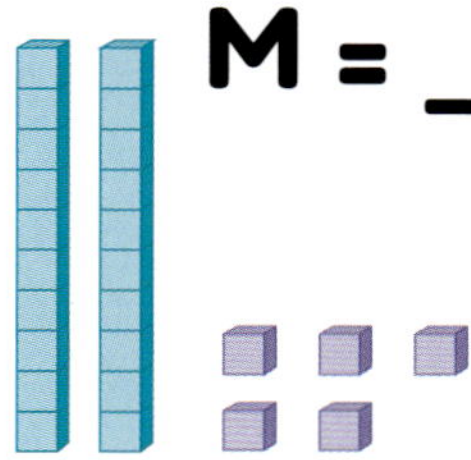 M = ___	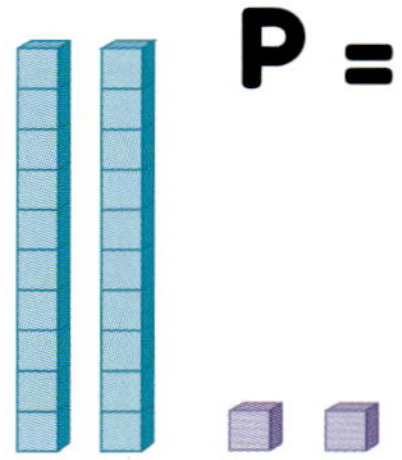 P = ___
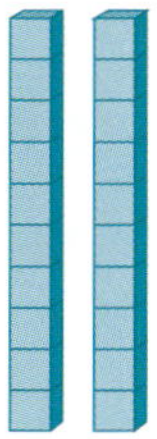 T = ___	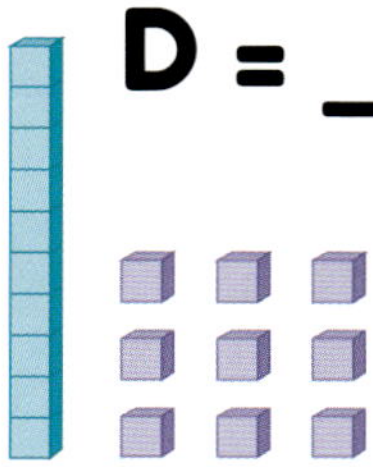 D = ___	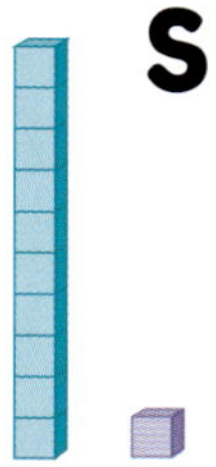 S = ___	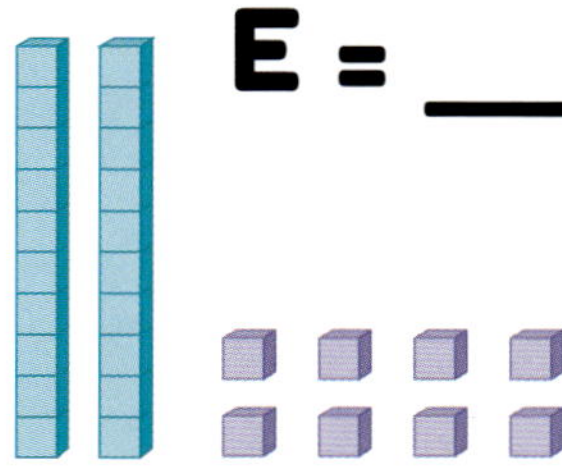 E = ___	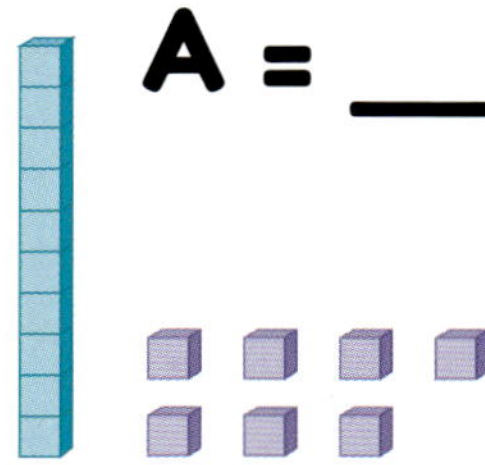 A = ___

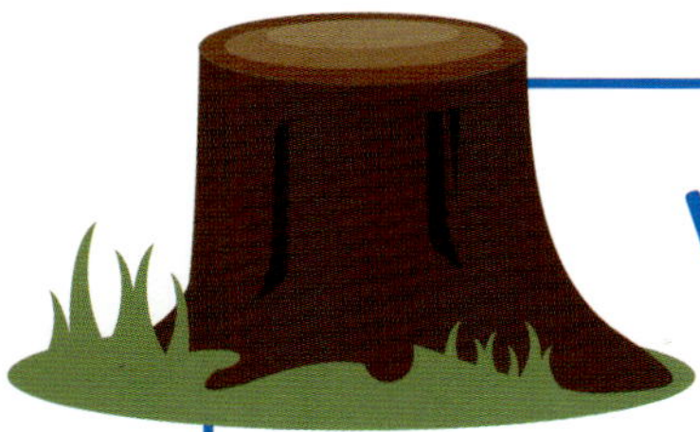

Why didn't the tree solve the math riddle?

___ ___   ___ ___ ___   ___ ___ ___ ___ ___ ___ ___!

12 20   16 17 11   11 20 4 25 22 28 19

Read

Kazoo and the Seed

Your child is ready to read the math-based storybook *Kazoo and the Seed*. This imaginative story includes visual representations of the place value chart as well as place value blocks (ones) and rods (tens). These math tools set the stage for your child to work with larger numbers consisting of three or more digits.

During reading:

Point out the place value rods and blocks on each page. Ask your child to count the rods and blocks to see if the total is the same as the number shown in the chart.

For a challenge, cover up the place value chart and ask your child to predict how the number should be written, based on what the place value blocks and rods show.

After reading:

Ask your child to draw their own Ten Tree, using a two-digit number. Which colors would represent the ones? The tens? How much would their tree grow in ten days?

Plant something with your child! Several seeds (or beans) placed loosely in a cup of soil should sprout fairly quickly.

Celebrate your child's progress by placing a star sticker in the workbook.

Value Roll

For each box, roll a die and write the number in the tens place. Roll again and write the number in the ones place. Then write the full number in the bottom space. An example has been done for you. Then answer the questions at the bottom of the page.

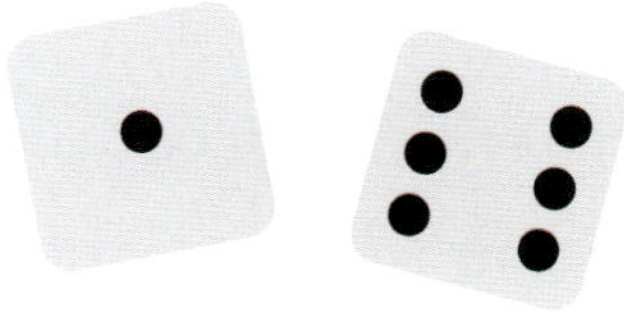

TENS	ONES
1	6
16	

Box 1

TENS ONES

Box 2

TENS ONES

Box 3

TENS ONES

Box 4

TENS ONES

Box 5

TENS ONES

1. What was the greatest number that you rolled? _____
2. What was the smallest number that you rolled? _____
3. Did you roll any doubles? ____ Yes ____ No
4. Were any two rolls exactly the same? ____ Yes ____ No
5. If you had to choose, which number would be your lucky number? _____

Your child has completed Step 8!
Celebrate by placing a star sticker in the workbook.

Skip Counting by 2s on a Number Line

Skip It

In math, *skip counting* means to count forward by a number that is not 1. To skip count, add the same number each time to the number that comes before it.

To skip count by 2s, add 2 to get to the next number. You can use a number line to help you skip count. When counting by 2s, you must count every other number each time.	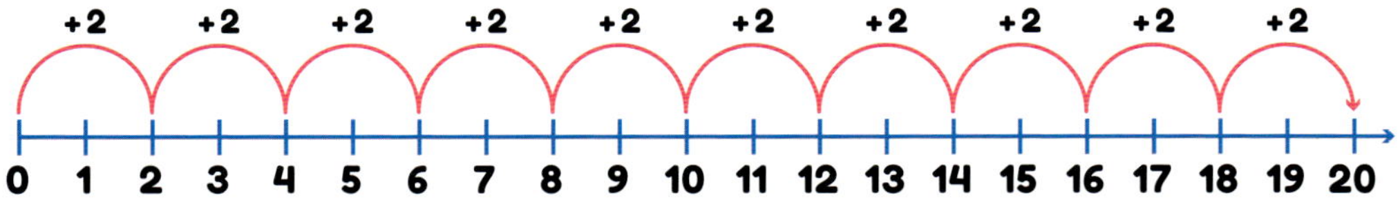
You can skip count by 2s from any number—you don't have to start at 0! Just add 2 each time to the number that comes before it.	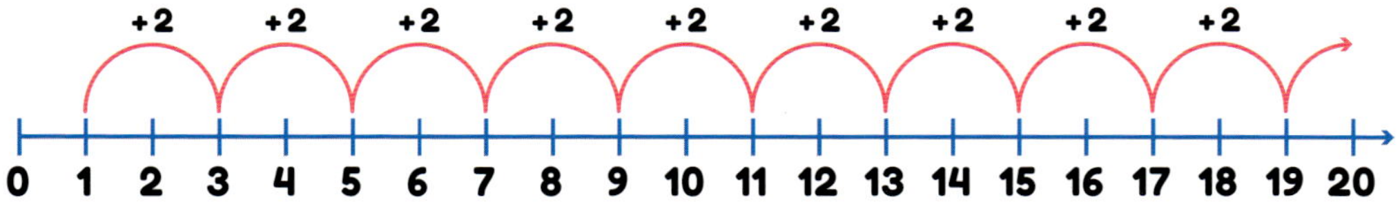

Skip to My Two!

Draw the skip counts by 2s from 4 to 14.

Draw the skip counts by 2s from 5 to 15.

Laundry Day

Look at each row of numbers. Fill in the missing numbers by skip counting by 2s.

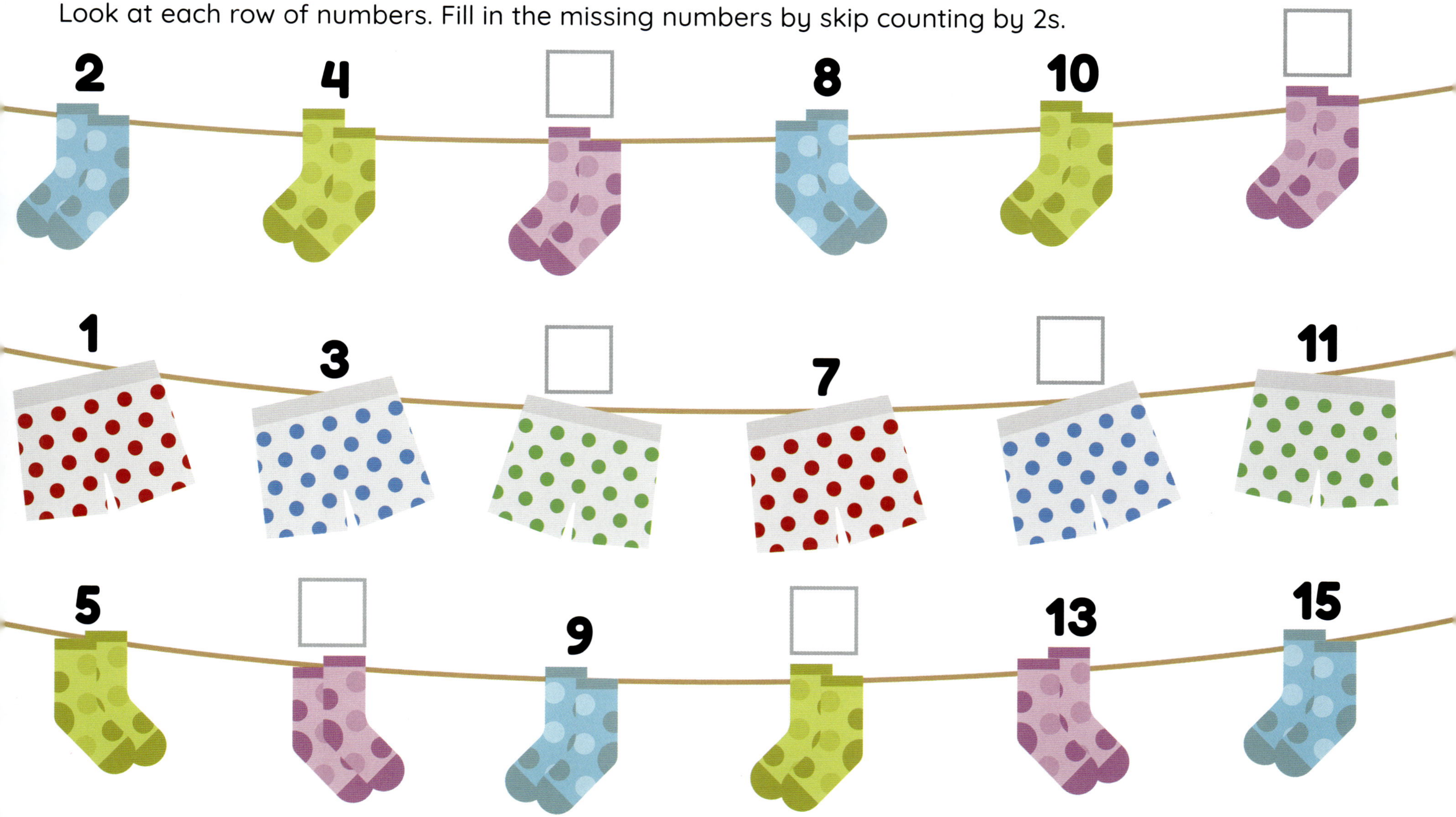

Skip Counting by 2s on a Hundred Chart

The hundred chart is another tool that can help you skip count.

The hundred chart does not start with 0. To skip count by 2s, start at **2** and go across the row: **2**, **4**, **6**, **8**, **10**.

Remember, you're still adding 2 to the number that comes before it.

+ 2 + 2 + 2 + 2

1	2	3	4	5	6	7	8	9	10
11	12	13	14	15	16	17	18	19	20
21	22	23	24	25	26	27	28	29	30
31	32	33	34	35	36	37	38	39	40
41	42	43	44	45	46	47	48	49	50
51	52	53	54	55	56	57	58	59	60
61	62	63	64	65	66	67	68	69	70
71	72	73	74	75	76	77	78	79	80
81	82	83	84	85	86	87	88	89	90
91	92	93	94	95	96	97	98	99	100

Notice that as you skip count across each row starting at **2**, the digits in the ones place repeat the same pattern: **2**, **4**, **6**, **8**, **0**.

Notice that as you skip count across each row starting at **1**, the digits in the ones place repeat the same pattern: **1**, **3**, **5**, **7**, **9**.

Use the hundred chart to skip count by 2s, starting with the first number on each caterpillar.

Odd and even numbers will be covered in future lessons.

Pearl Maze

Help Batu reach the giant pearl. Skip count by 2s and color in the numbers 2 through 50. You can use the hundred chart to help you.

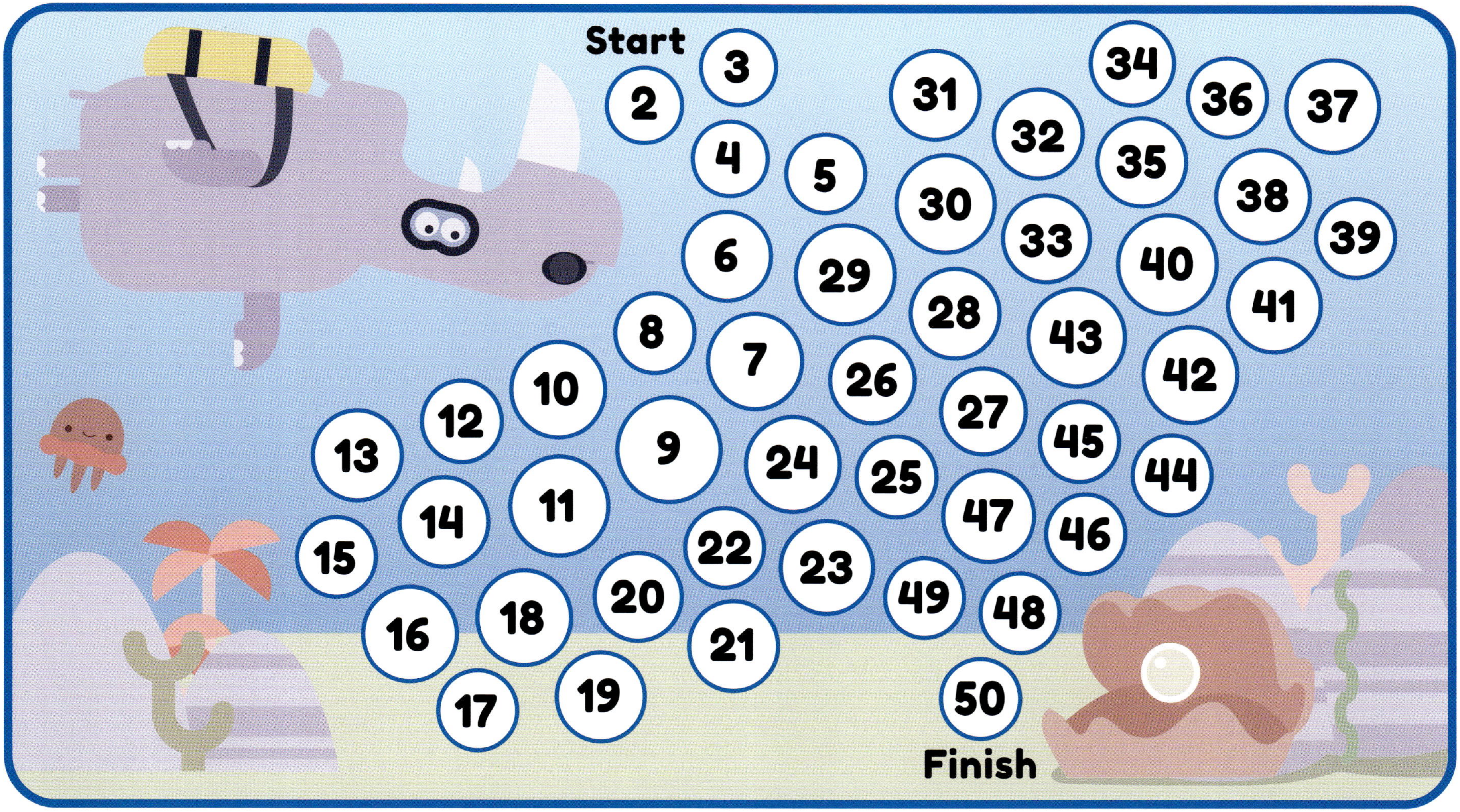

Skip Counting by 5s

Skip counting means to count forward by a number that is not 1. To skip count by 5s, add 5 to each number to get to the next number.

To skip count by 5s on a hundred chart, start at **5**. Add 5 to skip over the numbers to **10**. Then add 5 more to move to **15**, and then **20**.

Skip counting by 5s is a faster way to count groups of objects. Try it!

+5

1	2	3	4	5	6	7	8	9	10
11	12	13	14	15	16	17	18	19	20
21	22	23	24	25	26	27	28	29	30
31	32	33	34	35	36	37	38	39	40
41	42	43	44	45	46	47	48	49	50
51	52	53	54	55	56	57	58	59	60
61	62	63	64	65	66	67	68	69	70
71	72	73	74	75	76	77	78	79	80
81	82	83	84	85	86	87	88	89	90
91	92	93	94	95	96	97	98	99	100

Keep going to see a pattern. In the 5s column, the digit in the ones place is always 5, and in the tens column the last digit is always 0.

How many fingers are on the gloves below? Fill in the missing numbers by counting by 5s.

Bunches of Balloons

Help decorate for a party. Skip count by 5s and fill in the missing numbers to reach 100.

5	10	___	20	25
___	35	___	45	50
___	60	___	70	75
___	___	90	___	100

Launch

Skip Counting by 10s

Skip counting means to count forward by a number that is not 1. To skip count by 10s, add 10 to each number to get to the next number. This pattern happens for every column on the chart—take a look!

To skip count by 10s on a hundred chart, start at **10** and keep adding 10: **10**, **20**, **30**, **40**, **50**, **60**, **70**, **80**, **90**, **100**!

1	2	3	4	5	6	7	8	9	**10**
11	12	13	14	15	16	17	18	19	**20**
21	22	23	24	25	26	27	28	29	**30**
31	32	33	34	35	36	37	38	39	**40**
41	42	43	44	45	46	47	48	49	**50**
51	52	53	54	55	56	57	58	59	**60**
61	62	63	64	65	66	67	68	69	**70**
71	72	73	74	75	76	77	78	79	**80**
81	82	83	84	85	86	87	88	89	**90**
91	92	93	94	95	96	97	98	99	**100**

Keep going to see a pattern. The digit in the ones place does not change. The digit in the tens place goes up by one: 1, 2, 3, 4, 5, 6, 7, 8, and 9 until you reach **100**!

Use the hundred chart to skip count by 10s, starting with the first number on each caterpillar.

30 ___ ___ ___ ___ ___

34 ___ ___ ___ ___ ___

What's the Buzz?

How many buzzing bees? Count the bees by tens, then fill in the correct numbers on the flowers. Use the hundred chart to help you.

Under the Sea

Help the meerkat explore the ocean. Count by 2s to connect the dots. Then color the picture.

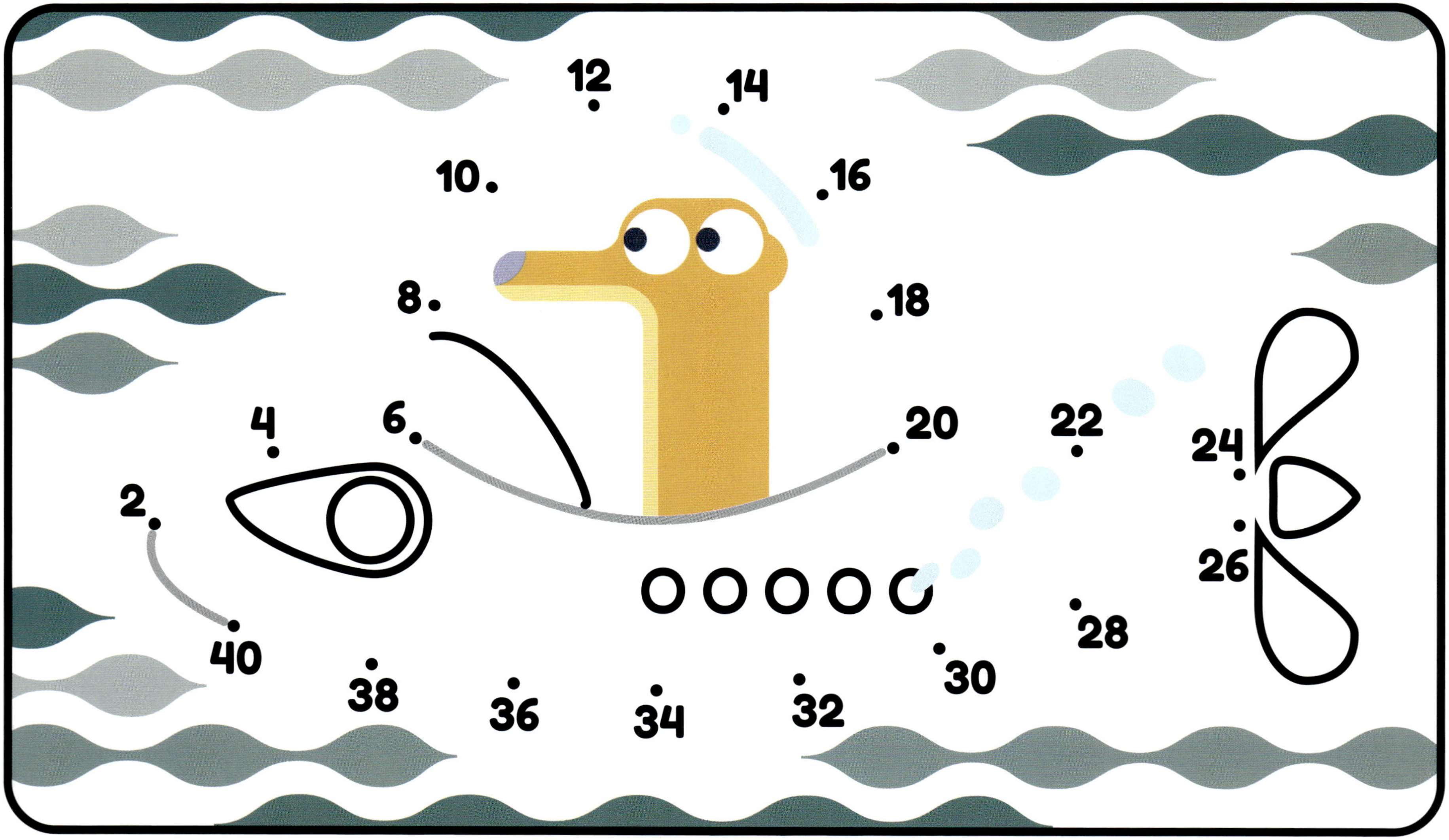

Example

Art Gallery

Math can be creative. Draw 10 dots or 10 of any small shape you like in each picture frame. Use any color and make any pattern you like. When you finish, skip count by tens, writing each number, until you reach 100!

Read 100 Fingers for 100 Days

Your child is ready to read the storybook *100 Fingers for 100 Days*, which celebrates the 100th day of school. The book features a series of hundred charts to help your child visualize skip counting by fives. The hundred chart is an essential tool for expanding your child's fluency with math facts.

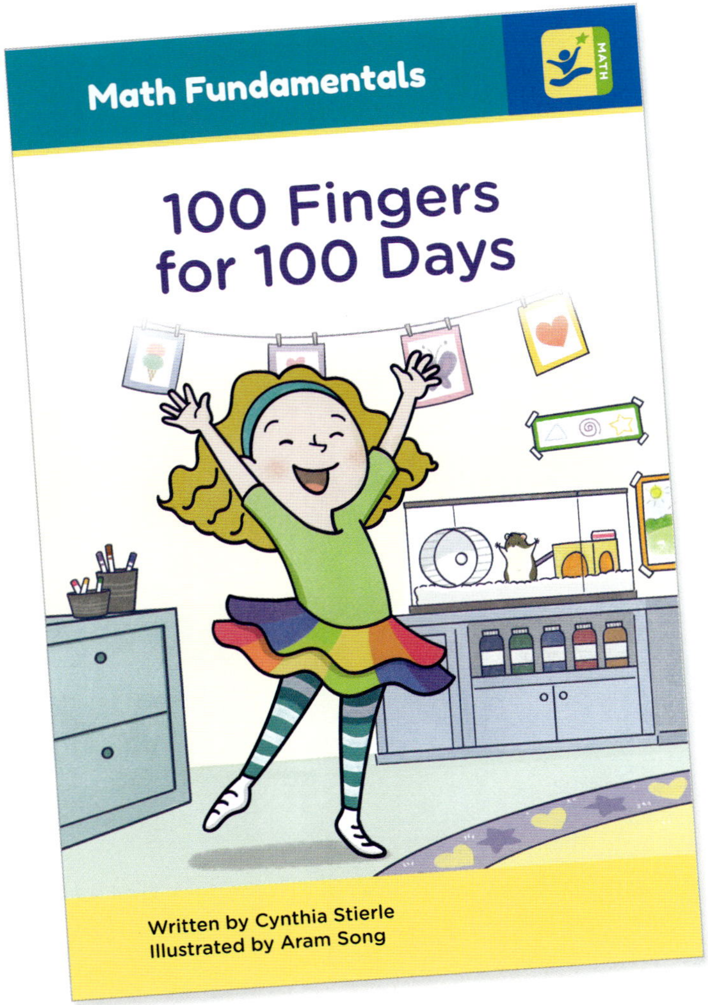

During reading:

Help your child recognize that the numbers on the hundred chart always follow a pattern. You may point out different patterns, such as skip counting by 2s and 10s.

After reading:

Make your own collection of 100 things. You can use pennies, buttons, marbles, or any other small object. Encourage your child to separate the collection into groups of 10, groups of 5, and even groups of 2.

How does your child's class celebrate the 100th day of school? Ask your child to tell you a story about their own school traditions for this special day.

Celebrate your child's progress by placing a star sticker in the workbook.

Fly Away

Get ready to take off on your next adventure. Fill in the missing numbers in each row by skip counting by 2s, 5s, and 10s. How high can you fly?

2	4			10			16		

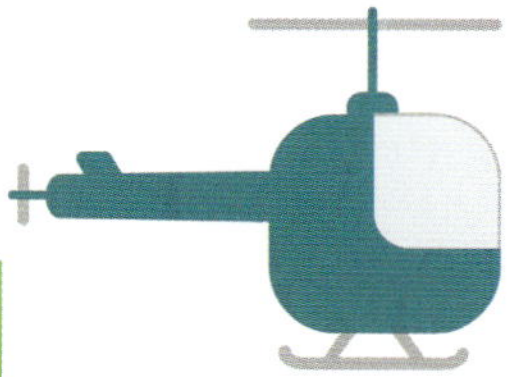

5			20		30			45	

10		30			60			90	

Bonus: What two numbers appear in all 3 rows? ☐ and ☐

HOOKED ON MATH

It's Time to Celebrate!

has completed Math Fundamentals Steps 8 & 9
and is now ready to move on to the
Math Fundamentals Steps 10 & 11 Workbook!

SKILLS MASTERED

- Working with a hundred chart
- Ordering three-digit numbers
- Place value: tens and ones
- Skip counting by 2s, 5s, and 10s

Don't forget to share your success and tag us on social media so we can celebrate with you!

Credits

Design Team
Julie Linden Frazier
Lindsey Frazier

Editorial Team
Donna Giachetti
Cynthia Stierle
Austin Feld

Illustrations
Dubit Limited

Educational Advisor
Lesley Freed, Curriculum Specialist

Published by Sandviks HOP, Inc.

Printed in China

HOOKED ON MATH

Math Fundamentals

18028 (07/22)

HOOKED ON MATH

Math Fundamentals

Steps 10 & 11

Workbook

Tips for Success

- Math Fundamentals introduces your child to the building blocks of math—number sense, addition, and subtraction.
- Work at a pace that's right for you and your child.
- Fit your session length to your child's attention span. It's okay to work with the App one day and this workbook the next.
- Before beginning, flip through the workbook and note any materials you might need. You should plan to have pencils, a pair of dice, and crayons ready.
- You'll see parent tips throughout this workbook. These tips are designed to answer questions you or your child might have about the learning.
- Repetition and review are essential to basic math fluency. Each workbook step builds on the ones before, so make sure your child has mastered the current lesson before moving forward.
- As you go through the workbook, encourage your child to explore the connection between numbers and the real world. Reinforce concepts with familiar household objects such as toys or snacks to make adding and subtracting more memorable.
- Have fun with it!

HOOKED ON MATH

Math Fundamentals

Steps 10 & 11

Workbook

This workbook and the included storybooks introduce fundamental math skills.

Step 10: Place Value
- Estimation
- Representing numbers
- Place value blocks to 120
- Comparing numbers to 120

Step 11: Two-digit Addition and Subtraction
- Addition and subtraction with place value blocks
- Addition and subtraction with place value charts

Contents

Lesson Overview

Welcome! This workbook provides supplemental activities that will solidify and expand the learning your child is enjoying with the Hooked on Math App. Follow the easy, step-by-step instructions to guide your child's exploration of numbers and operations. Every lesson helps your child build a strong foundation in fundamental math skills while having fun!

1. Launch

Your child will be introduced to math concepts and skills through interactive lessons in the App on your mobile device or Windows or Mac computer. Download the App at:
hookedonmath.com/get-the-app

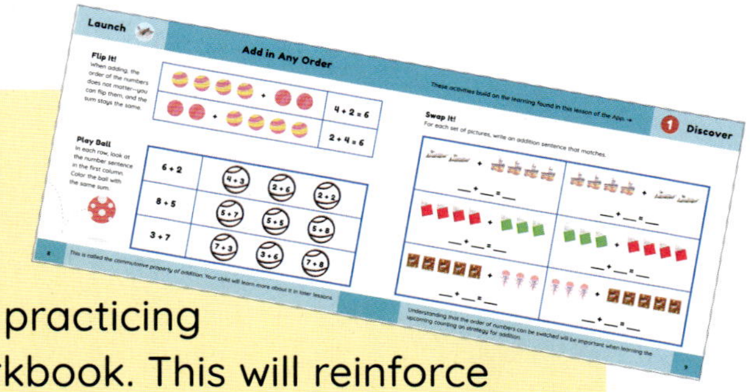

2. Discover

Next, your child will have fun exploring math concepts and practicing new skills in the workbook. This will reinforce the App lessons and provide a tangible application of the skills learned.

3. Play

In addition to the workbook activities, your child will enjoy two math-based storybooks that will strengthen their learning and add to the fun!

4. Celebrate

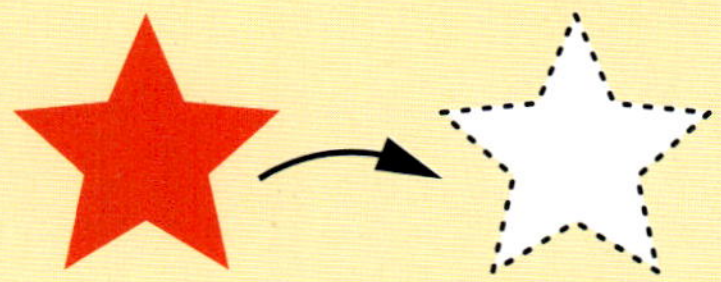

Celebrate your child's progress by placing a sticker in the workbook as learning milestones are reached. This also provides a helpful marker to remember where to begin next time.

Special notes are included in this space to provide additional context related to what is being taught.

Hundred Chart Mystery Picture

Use the Color Key to color in the squares and reveal a mystery picture.

Color Key

Black: 4, 7, 15, 16, 25, 26, 35, 36, 45, 46, 55, 56, 65, 66, 75, 76, 85, 86, 95, 96

Orange: 22, 23, 28, 29, 33, 38, 53, 58, 62, 63, 68, 69, 72, 73, 78, 79, 83, 88

Purple: 32, 39, 82, 89

Blue: 1, 2, 3, 5, 6, 8, 9, 10, 11, 14, 17, 20, 41, 50, 51, 60, 91, 94, 97, 100

Yellow: all remaining numbers

1	2	3	4	5	6	7	8	9	10
11	12	13	14	15	16	17	18	19	20
21	22	23	24	25	26	27	28	29	30
31	32	33	34	35	36	37	38	39	40
41	42	43	44	45	46	47	48	49	50
51	52	53	54	55	56	57	58	59	60
61	62	63	64	65	66	67	68	69	70
71	72	73	74	75	76	77	78	79	80
81	82	83	84	85	86	87	88	89	90
91	92	93	94	95	96	97	98	99	100

Materials needed: crayons or colored pencils in black, orange, purple, blue, and yellow

Tow a Row

In each row, look at the number on the tugboat. Color enough rods and blocks to match that number. Then write the digits in the place value chart. The first one has been done for you.

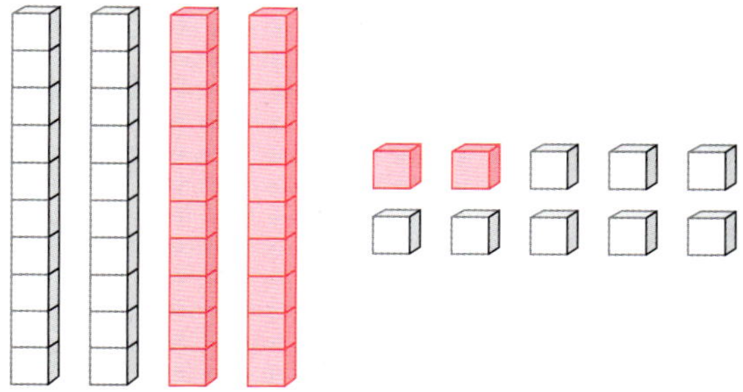

Tens	Ones
2	2

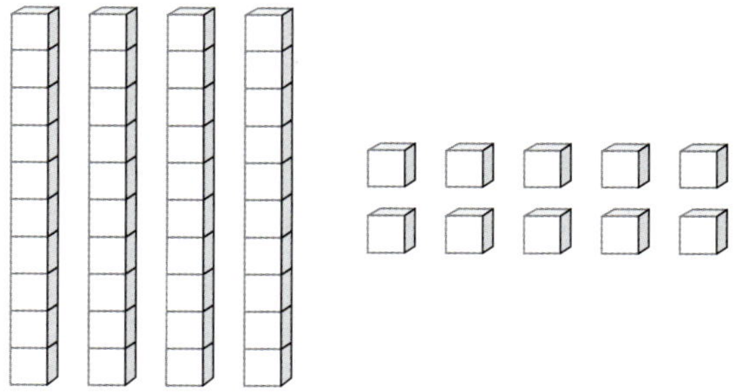

Tens	Ones

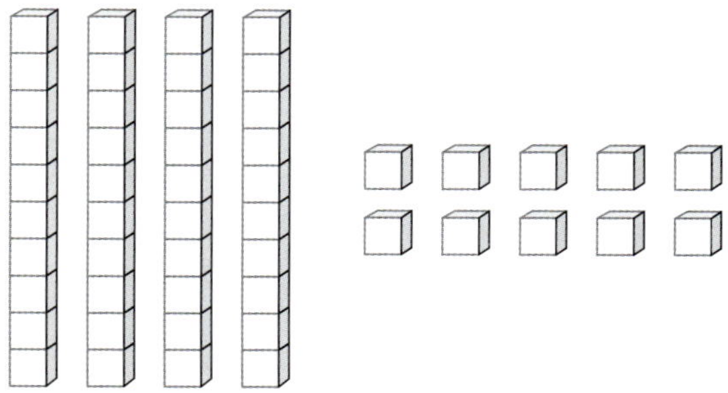

Tens	Ones

Goody Two Shoes

Skip count by 2s. Write the missing number above each pair of shoes.

2	4	☐	8	☐	12

14	☐	18	20	22	☐

☐	28	☐	32	☐	36

Skip Counting by 10s

Review

Quick Count

Count by groups of 10 and write the total in each box.

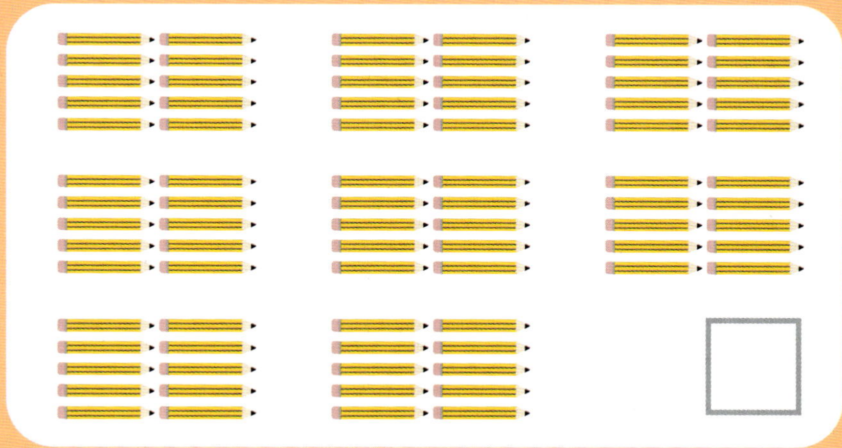

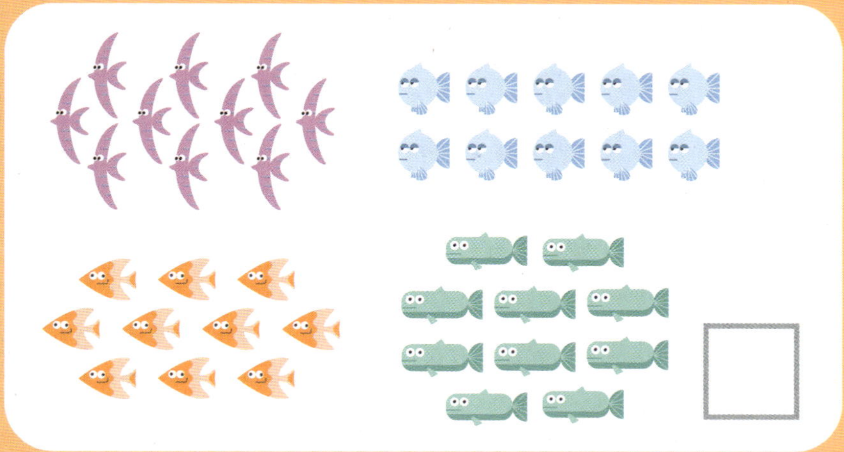

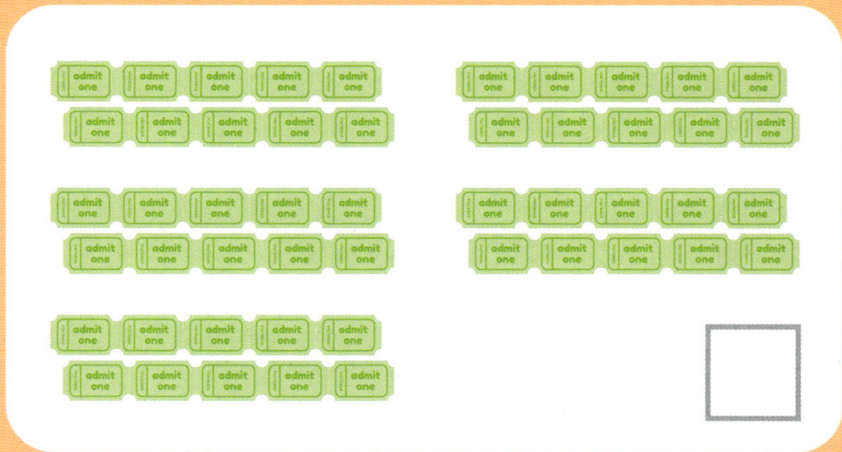

Celebrate finishing the review with a star sticker.

Estimation

In math, guessing is called *estimating*. An estimate is not an exact number, but is a thoughtful guess based on the information that you know.

One way to estimate is to try to group objects in sets of 10 without counting, then guess how many sets of ten there are in the entire group. Without counting, you can guess there are about 10 apples in each group below. Skip count by 10s to reach 40.

A guess of 40 is close to the actual number—37.

Math Talk
When estimating numbers, you might use words such as *almost*, *about*, *close enough*, or *a little more (or less) than*.

Rain, Rain

How many raindrops all together? Circle the closest estimate. Then count the raindrops to check your answer.

About 40 **About 60** **About 100**

Estimating is a real-world skill that takes practice. Your child may find it difficult to "guess" instead of providing an exact answer. Remind them that sometimes it's okay to be "close enough."

Close Enough

Estimate the number of ocean pals in the scene below. Write your guess. Then count the actual number of pals. The first one has been done for you.

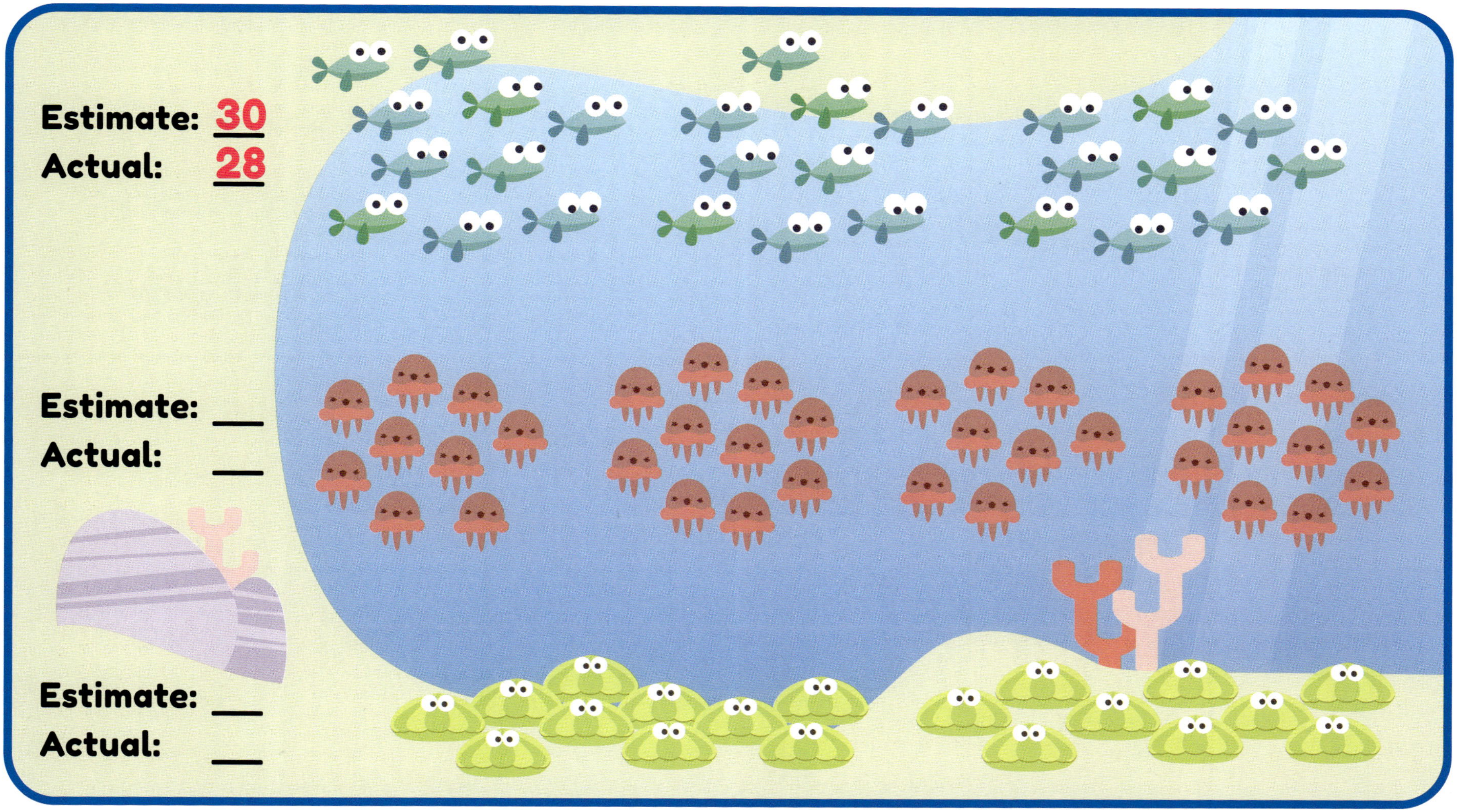

As your child progresses to working with larger numbers, estimating the answer can help them identify errors.

Representing Numbers

Numbers can be shown in many ways. For example:

Pictures and objects

Digits

1 2 3

4 5 6

Words

one

three

twenty-one

forty-two

Place value chart

A place value chart shows the value of each digit in any number. The value depends on the location of each digit.

Tens	Ones
2	1

Number sentences

Horizontal number sentences

60 + 10 = 70

80 – 10 = 70

Vertical number sentences

$$\begin{array}{r} 60 \\ +\,10 \\ \hline 70 \end{array} \qquad \begin{array}{r} 80 \\ -\,10 \\ \hline 70 \end{array}$$

Tools

Tools like these can represent numbers.

Number line 

Ten frame 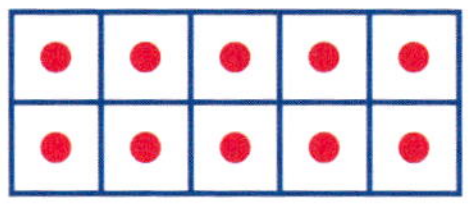

Hundred chart 

A *hyphen* looks like a small dash or line and joins two words together. When writing a number in word form, include a hyphen between the number words, starting with number twenty-one.

Represent!

Experiment with different ways to represent the same number.

- Trace the numeral.
- Fill in the ten frame.
- Mark the number on the number line.
- Trace the number word.
- Circle the number on the hundred chart.
- Color the correct number of tens rods and ones blocks.

17 17 17

seventeen

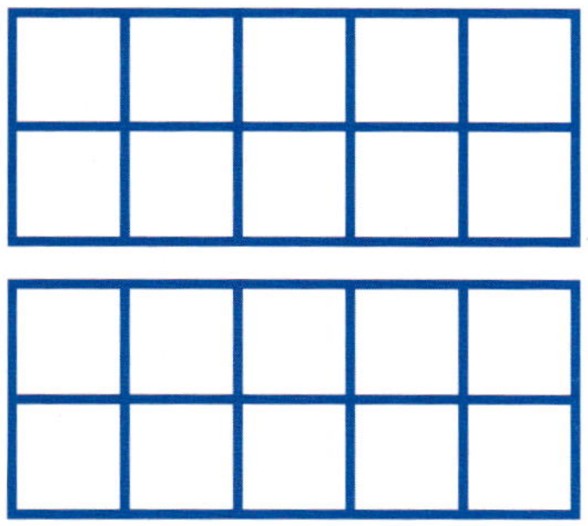

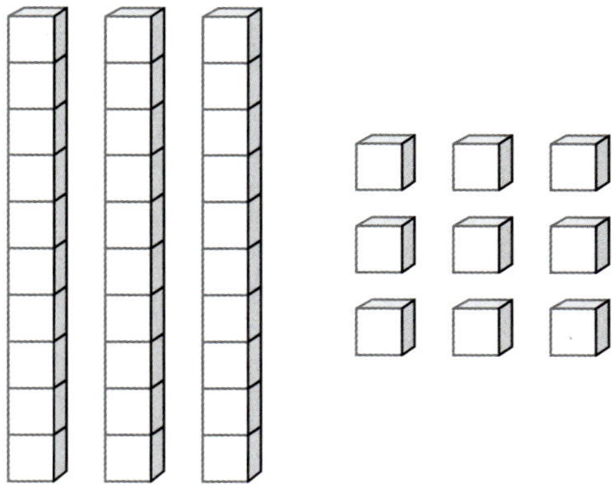

1	2	3	4	5	6	7	8	9	10
11	12	13	14	15	16	17	18	19	20
21	22	23	24	25	26	27	28	29	30
31	32	33	34	35	36	37	38	39	40
41	42	43	44	45	46	47	48	49	50
51	52	53	54	55	56	57	58	59	60
61	62	63	64	65	66	67	68	69	70
71	72	73	74	75	76	77	78	79	80
81	82	83	84	85	86	87	88	89	90
91	92	93	94	95	96	97	98	99	100

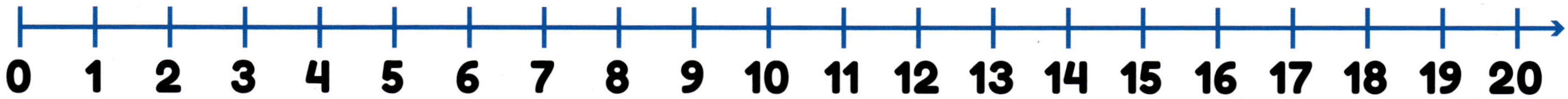

Place Value Blocks to 50

A number can be shown with place value blocks and rods.
- Each rod is equal to 10.
- Each block is equal to 1.

Tens	Ones
4	9

STOP!

Review the place value rules.
- There can only be one digit in each place.
- The digit in the tens place represents groups of 10.

GO!

Look at the rods and blocks below. Write the number.

Tens	Ones	Write It!
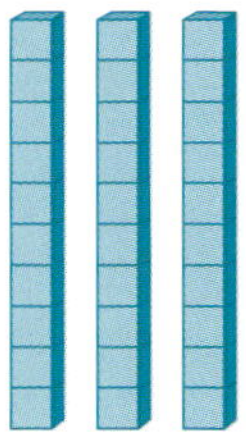		___
	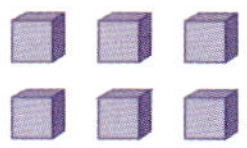	___

It is important for your child to understand what each place represents in a place value chart before moving on to addition and subtraction problems that involve regrouping.

In the Clouds

Look at the sets of rods and blocks in the clouds. Write the number for each set. Then draw the greater than (**>**) or less than (**<**) symbol in the circle to compare the numbers.

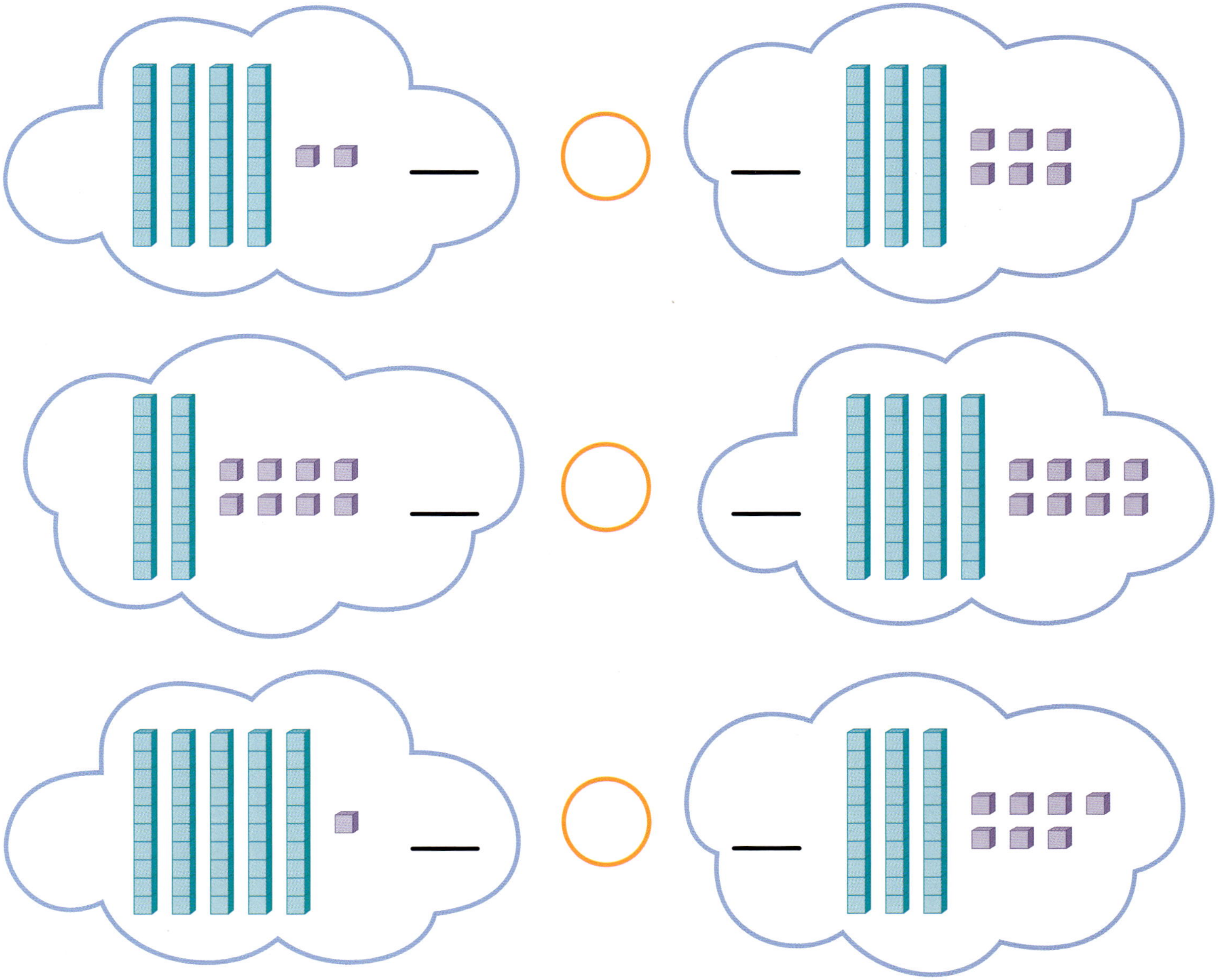

Place Value Blocks to 120

Make 100

You know that 10 ones blocks are equal to 1 tens rod.

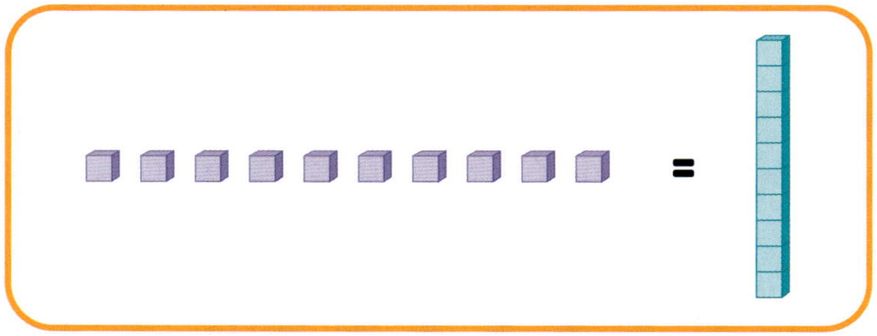

But what about 10 tens rods? They equal 100, or a hundreds *flat*.

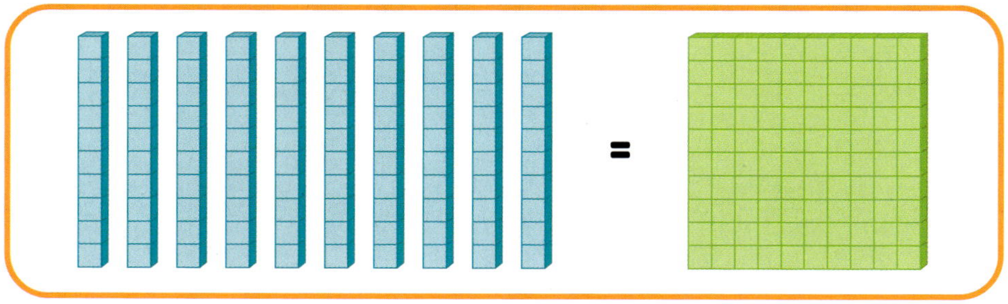

Notice that the hundreds flat looks like a hundred chart!

Three-digit Numbers

There are three digits in the number 100. When working with three-digit numbers, another place value position is added. The hundreds place is next to the tens place. The digit in the hundreds place represents a group of one hundred. But if there are no hundreds, then no 0 is needed, as shown in the number 99.

The number 100, when written in digits, has a 1 in the hundreds place, a 0 in the tens place, and a 0 in the ones place.

Hundreds	Tens	Ones
1	0	0
	9	9

Put It in Place

Look at the place value flats, rods, and blocks in each row. Write the number of hundreds, tens, and ones in the place value chart. Then write the number on the line. The first one has been done for you.

Hundreds	Tens	Ones
1	1	3

113

Hundreds	Tens	Ones

____

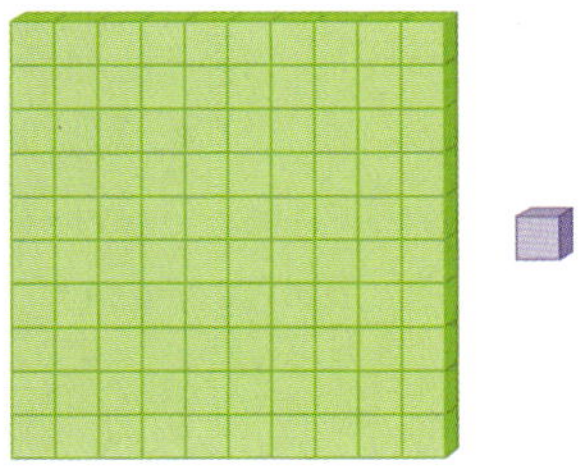

Hundreds	Tens	Ones

____

Zero at Work: See how 0 works as a placeholder? If there is a hundreds flat, but no tens rods or ones blocks, write a 0 in the place value chart for tens or ones.

Comparing Numbers to 120

Pairs to Compare

Use a place value chart to compare three-digit numbers.

	Hundreds	Tens	Ones		Hundreds	Tens	Ones
Start by comparing the digits in the greatest place value, the hundreds.	1	2	0	>		9	9
If the digits in the hundreds place are the same, move to the next greatest place value, the tens.	1	2	0	>	1	1	9
If the digits in the hundreds and the tens place are the same, move to the ones.	1	1	5	<	1	1	7

> greater than < less than = equal to, or the same

Color Key

\> < =

Catching Some Rays

Write the correct sign to make each number sentence true. Then color the manta ray next to the number sentence according to the Color Key. The first one has been done for you.

106 < 116

112 ◯ 21

93 ◯ 119

53 ◯ 53

86 ◯ 68

102 ◯ 120

Materials needed: crayons or colored pencils in orange, pink, and blue

Which Doesn't Belong?

The numbers below are represented in four different ways. Three of the numbers are the same in each row, but one is different. Find and circle it.

<table>
<tr><th>Place Value Blocks</th><th>Standard Form</th><th>Words</th><th>Other</th></tr>
<tr><td>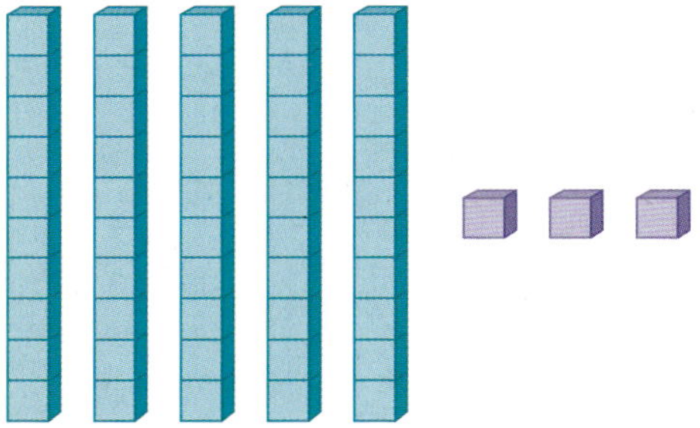</td><td>53</td><td>fifty-three</td><td>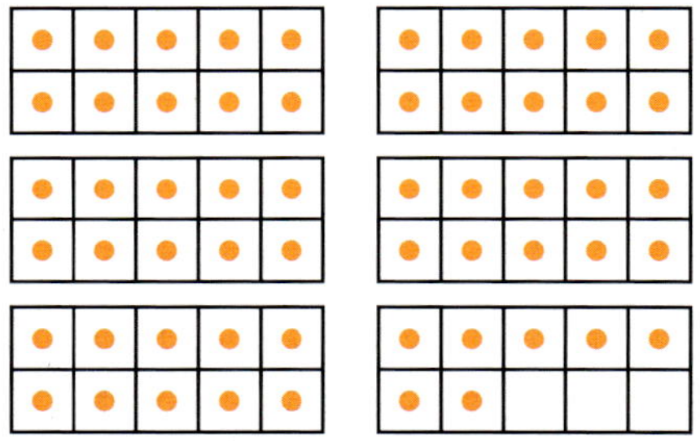</td></tr>
<tr><td>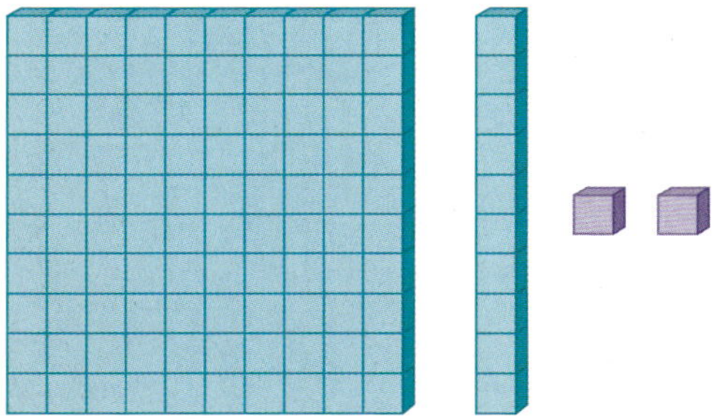</td><td>112</td><td>one hundred twelve</td><td>
</td></tr>
<tr><td></td><td>96</td><td>ninety-six</td><td>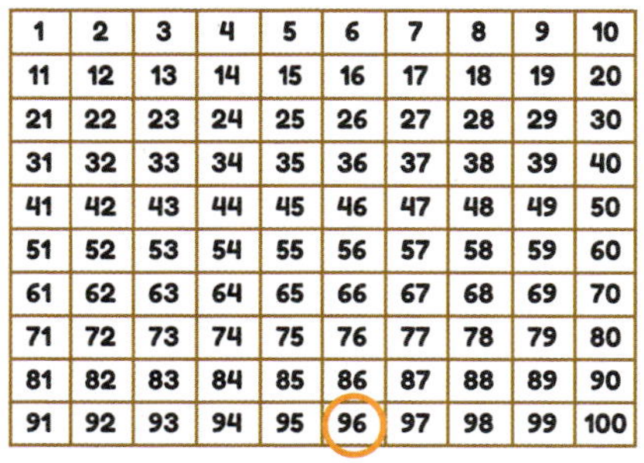
</td></tr>
</table>

Guessing Gumballs

Take a quick look at the gumballs and try to guess how many there are. Circle the number that is closest to your guess. Then count and color the gumballs to find the exact amount. Was your guess close?

The ability to estimate is an indication of a child's solid grasp of number sense and how numbers relate to each other. If your child looks at a group of objects and estimates far greater or fewer than shown, it may help to go back and review earlier lessons.

15

50

96

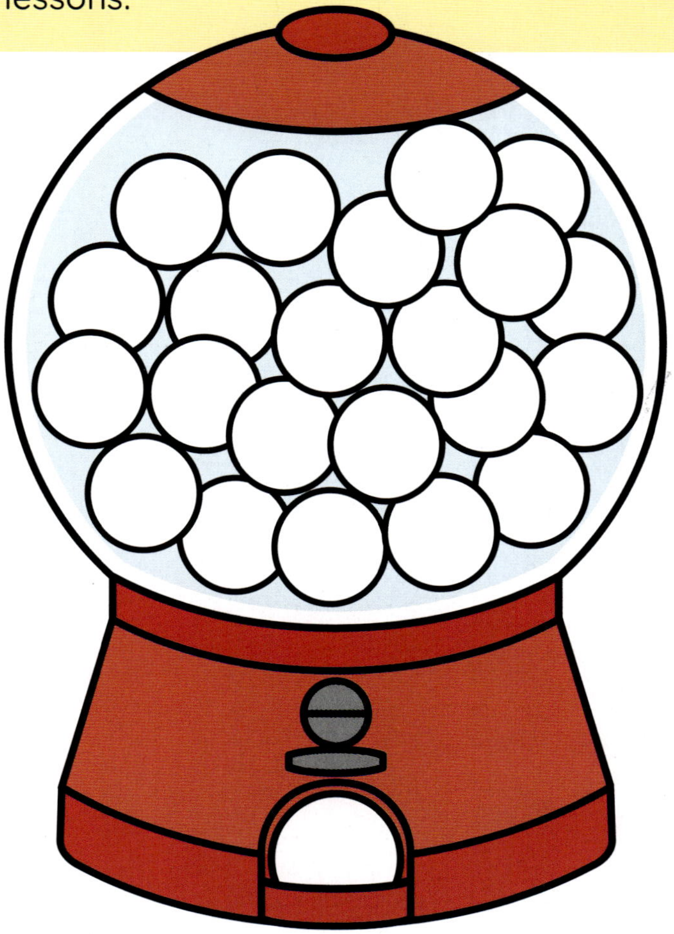

99

65

20

Read

The Seesaw

Your child is ready to read *The Seesaw*. This imaginative story uses a fictional measurement of weight called a "palump" to help reinforce your child's understanding of the concepts of *greater than* and *less than*.

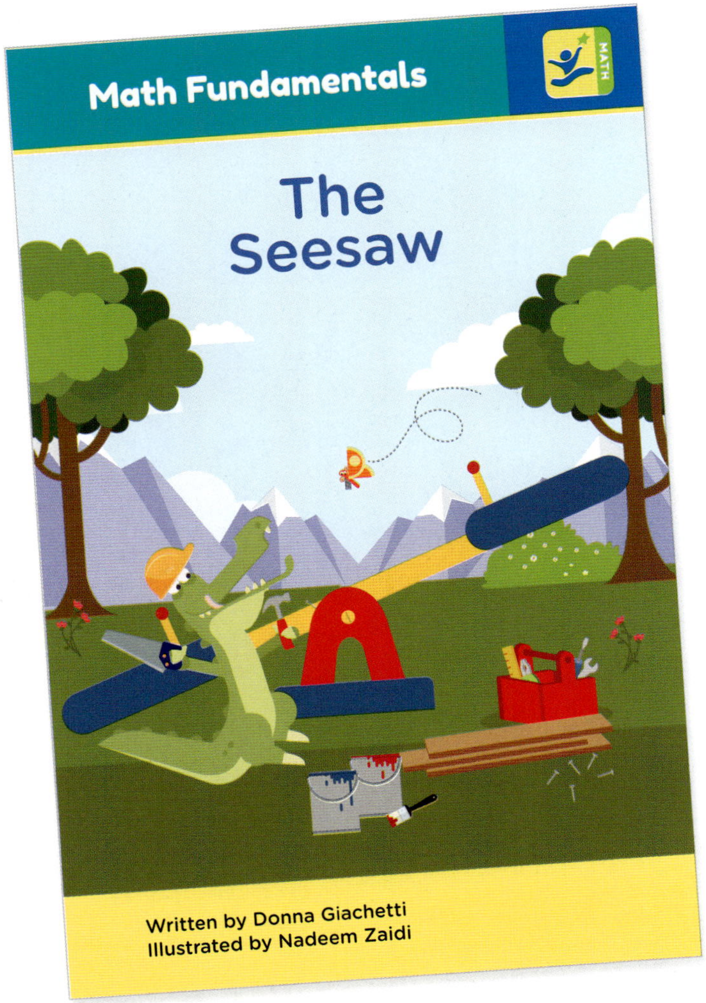

Before reading:

Point out the image of Chow on the inside front cover. Be sure your child understands that the text "80 palumps" refers to how much Chow weighs.

During reading:

See if your child can predict which side of the seesaw will go up, based on the characters on either side.

After reading:

If you have a scale available, let your child experiment with weighing different items. Choose two items and ask your child to predict whether one will weigh more than, less than, or the same as the other. Then weigh them to find out!

Invite your child to draw different seesaw options for the pals in the story. On the inside back cover you will find the specific weight (in palumps) of each animal.

Celebrate your child's progress by placing a star sticker in the workbook.

Place Value Block Code

Look at the place value flats, rods, and blocks in the Code Key. Write the number in each box. Then write the letter that matches the number in the spaces below to answer the riddle.

Code Key

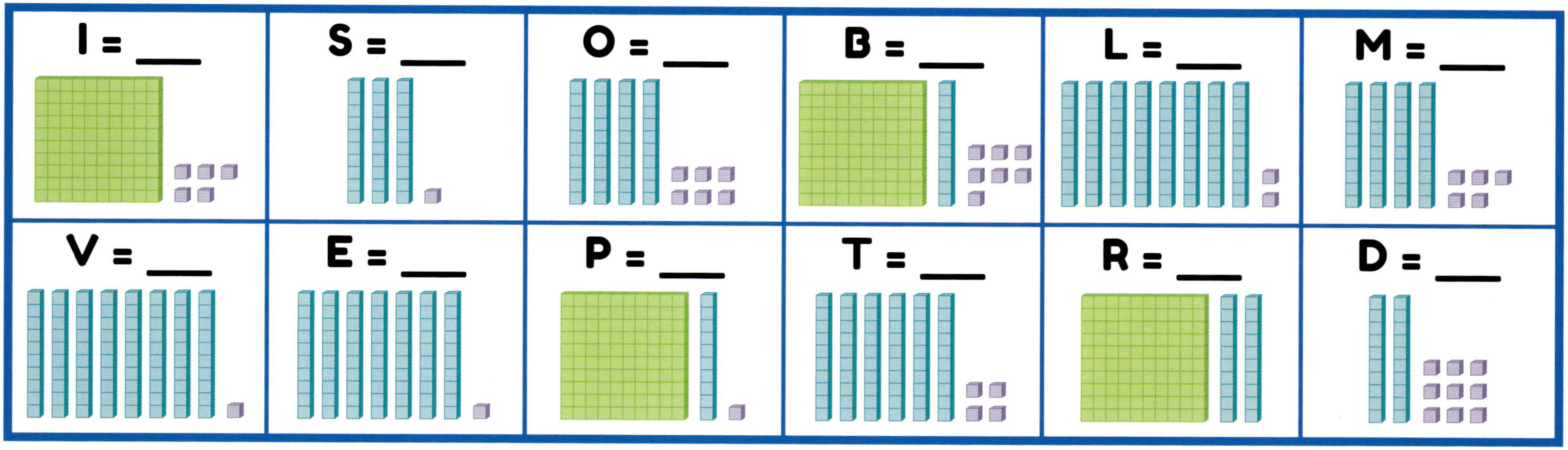

Why did the math book like kids?

Because they

___ ___ ___ ___ ___ ___   ___ ___ ___   ___ ___ ___ ___ ___ ___ ___ ___ !

31 46 82 81 71 29   105 64 31   111 120 46 117 82 71 45 31

Place Value Blocks Addition

Place value blocks can be used to solve double-digit addition problems.

Follow the steps below to solve the problem 32 + 23 = ?

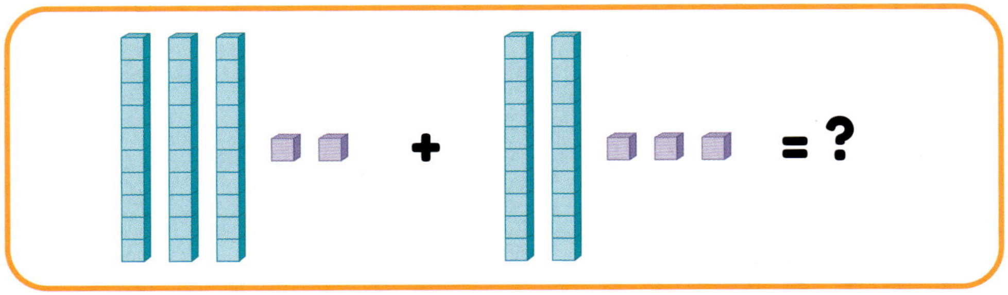

		Tens	Ones
Add the ones together. 2 + 3 = 5			5
Add the tens together. 30 + 20 = 50	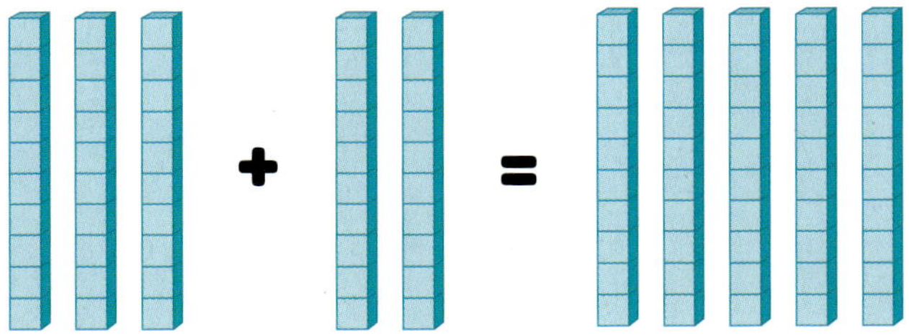	5	
Add the tens and ones together. 50 + 5 = 55	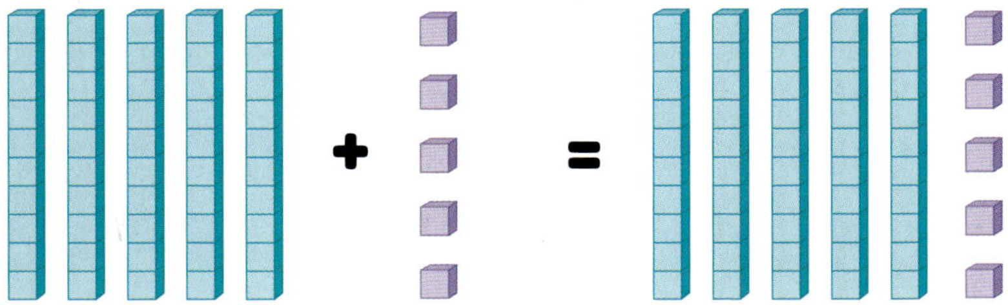	5	5

How many jellyfish?

On the Block

Look at each group of place value rods and blocks.

- Write the number under each set.
- Add the groups together and color the sum.
- Then write the numeral. The first one has been done for you.

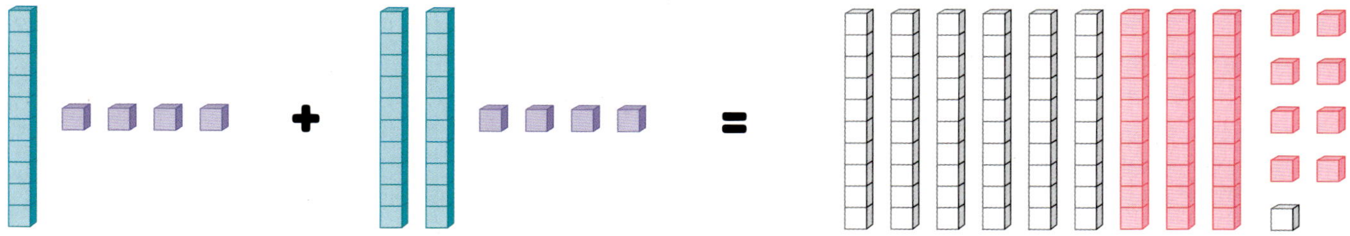

14 + 24 = 38

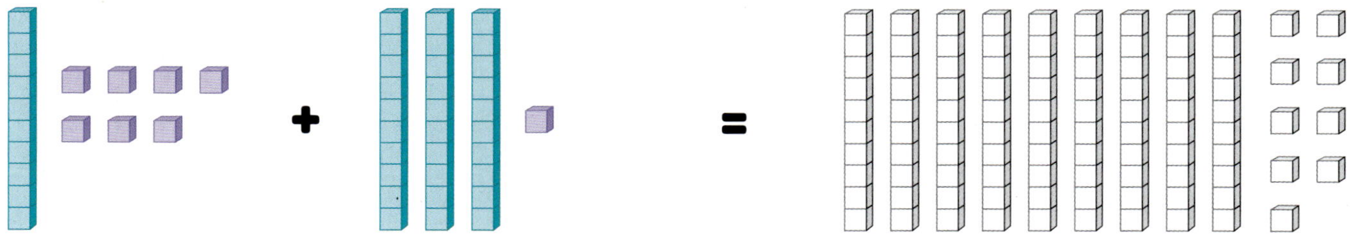

___ + ___ = ___

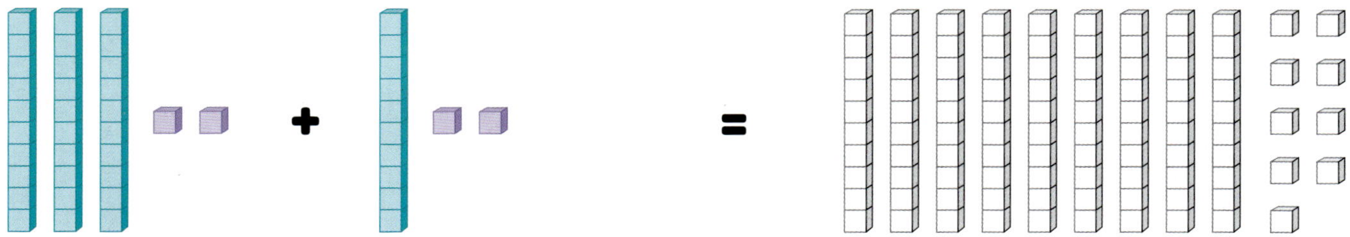

___ + ___ = ___

Launch

Place Value Blocks Subtraction

Place value rods and blocks can be used to solve double-digit subtraction problems.

Follow the steps below to solve the problem 45 - 33 = ?

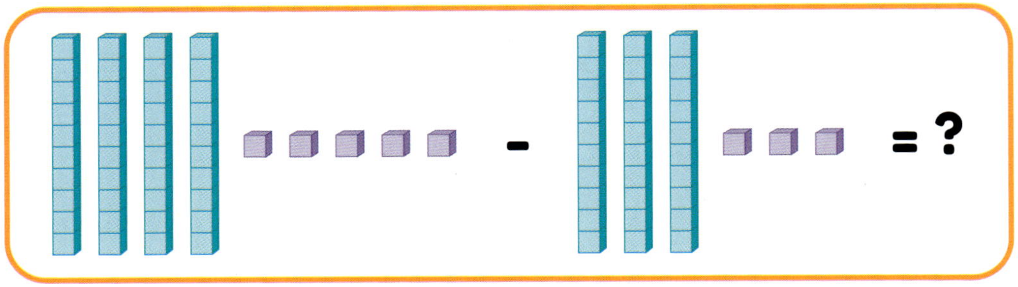

Subtract the ones. 5 - 3 = 2	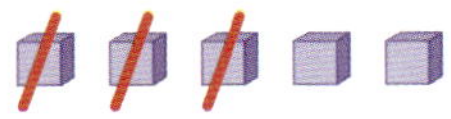	Tens: Ones: 2
Subtract the tens. 40 - 30 = 10	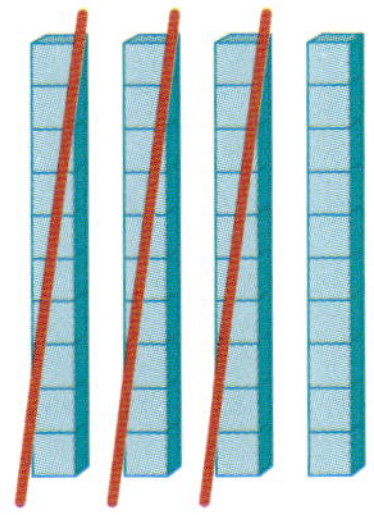	Tens: 1 Ones:
Put the remaining tens and ones together. 10 + 2 = 12	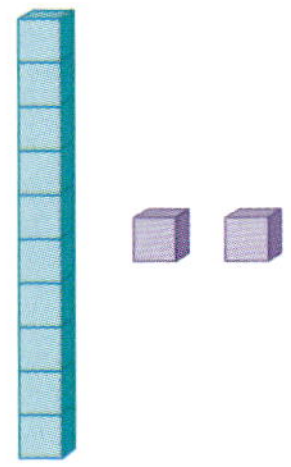	Tens: 1 Ones: 2

Block It Out

Look at each group of place value rods and blocks.

- Read the subtraction sentence.
- Cross out the number of rods and blocks being taken away.
- Then write the difference. The first one has been done for you.

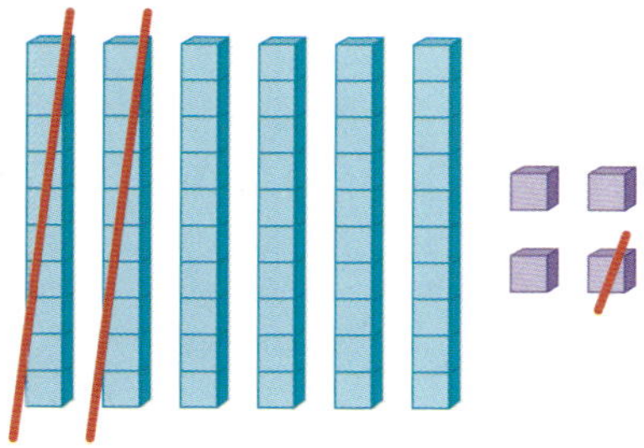

64 - 21 = 43

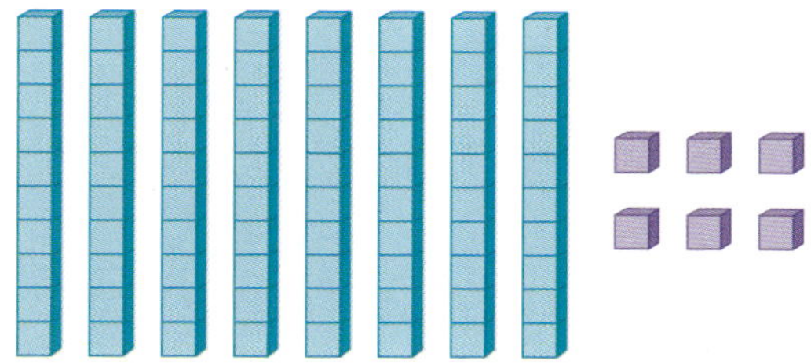

86 - 72 = ___

57 - 25 = ___

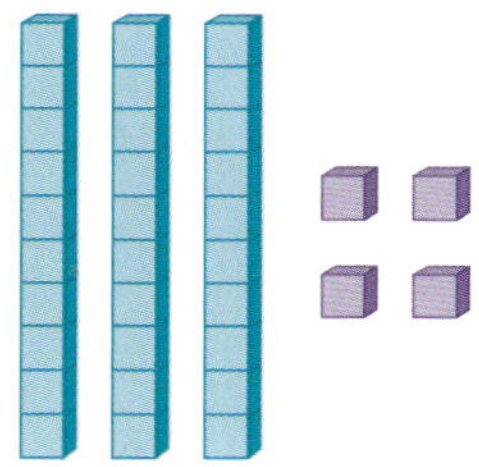

34 - 12 = ___

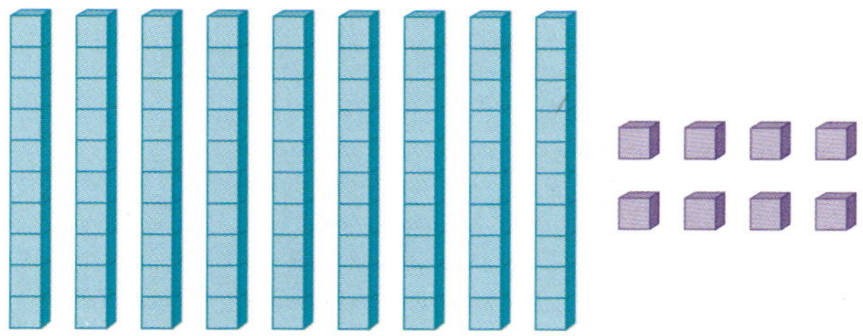

98 - 81 = ___

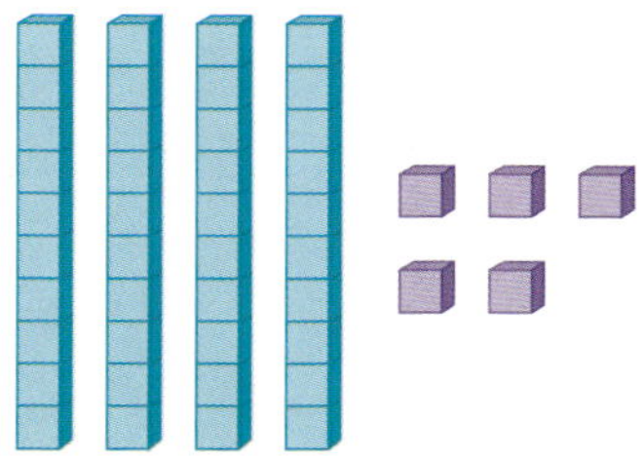

45 - 34 = ___

Tens and Ones Addition

You can add two-digit numbers in a place value chart.
64 + 32 + ?

Write the numbers so that the place values line up.

Tens	Ones
6	4
+ 3	2
9	6

First, add the numbers in the ones place: 4 + 2 = 6. Write a 6 in the ones column.

Next, add the numbers in the tens place: 6 + 3 = 9. Write a 9 in the tens column.

64 + 32 = 96

Add 'em Up

Add the two-digit numbers in Cake 1. Write the sum in the top row of Cake 2 and add those numbers. Write that sum in the top row of Cake 3, and the sum of Cake 3 in the top row of Cake 4. What's the final number?

Tens	Ones
2	2
+ 1	1

Cake 1

Tens	Ones
+ 2	0

Cake 2

Tens	Ones
+ 2	1

Cake 3

Tens	Ones
+ 2	5

Cake 4

Elevator Excitement

Kazoo pressed all the buttons in the elevator! Each pal will get out on a different floor, but which one? To find out, add the numbers together in each place value chart. Then write the sum in the chart and in the sentence.

Circle the pal that gets out on the highest floor of the skyscraper. The first one has been done for you.

Tens	Ones
4	3
+ 2	6
6	9

Gabby's floor is 69.

Tens	Ones
3	5
+ 1	3

Zola's floor is ____.

Tens	Ones
8	8
+ 1	1

Chow's floor is ____.

Tens	Ones
5	2
+ 4	1

Kip's floor is ____.

Tens	Ones
6	6
+ 2	0

Tali's floor is ____.

Tens and Ones Subtraction

You can subtract two-digit numbers in a place value chart.
76 - 45 = ?

Write the numbers so that the place values line up.

Tens	Ones
7	6
- 4	5
3	1

First, subtract the numbers in the ones place: 6 - 5 = 1. Write a 1 in the ones column.

Next, subtract the numbers in the tens place: 7 - 4 = 3. Write a 3 in the tens column.

76 - 45 = 31

Take 'em Down

Subtract the bottom number from the top number in Cake 1. Write the difference in the top row of Cake 2. Solve the subtraction problem and write that difference in the top row of Cake 3. Keep going until you reach the final number.

Tens	Ones
9	9
- 2	1

Cake 1

Tens	Ones
- 2	0

Cake 2

Tens	Ones
- 3	5

Cake 3

Tens	Ones
- 1	3

Cake 4

Elevator Exhaustion!

Kazoo pressed all elevator buttons again! Each pal will get out on a different floor, but which one? To find out, solve the subtraction problems in each place value chart. Then write the difference in the chart and in the sentence.

Circle the pal that gets out on the lowest floor of the skyscraper. The first one has been done for you.

Tens	Ones
8	5
- 4	2
4	3

Gabby's floor is 43.

Tens	Ones
9	9
- 6	5

Zola's floor is ____.

Tens	Ones
7	6
- 3	6

Chow's floor is ____.

Tens	Ones
5	2
- 4	0

Kip's floor is ____.

Tens	Ones
4	6
- 4	5

Tali's floor is ____.

More or Less

- Solve the addition problems on both sides of the seesaw.
- Write each sum as a numeral inside the triangle.
- Then draw the correct math symbol in the circle (**>**, **<**, **=**) to make the math sentence true. The first one has been done for you.

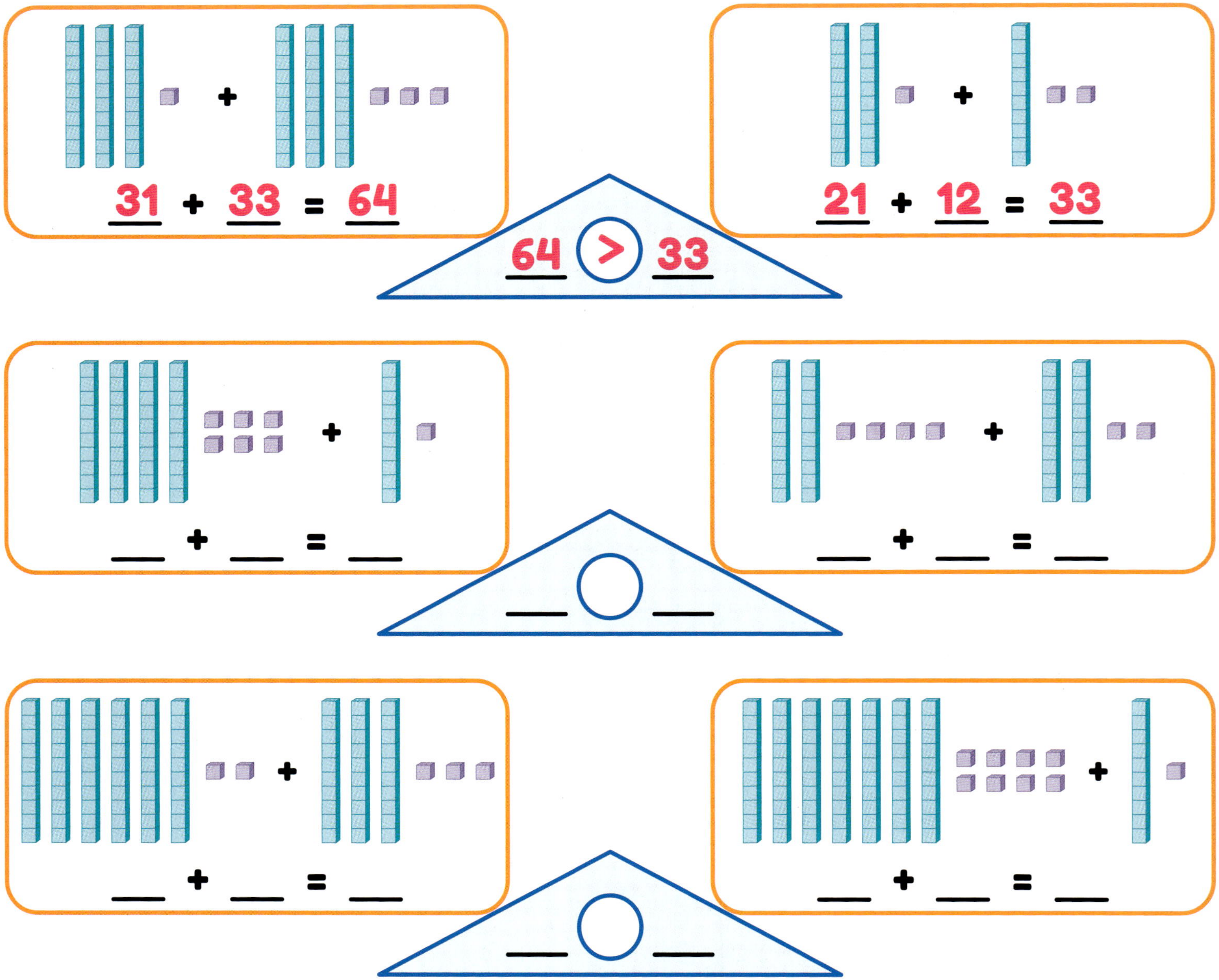

Less or More

- Solve the subtraction problems on both sides of the seesaw.
- Write each difference as a numeral inside the triangle.
- Then draw the correct math symbol in the circle (**>**, **<**, **=**) to make the math sentence true. The first one has been done for you.

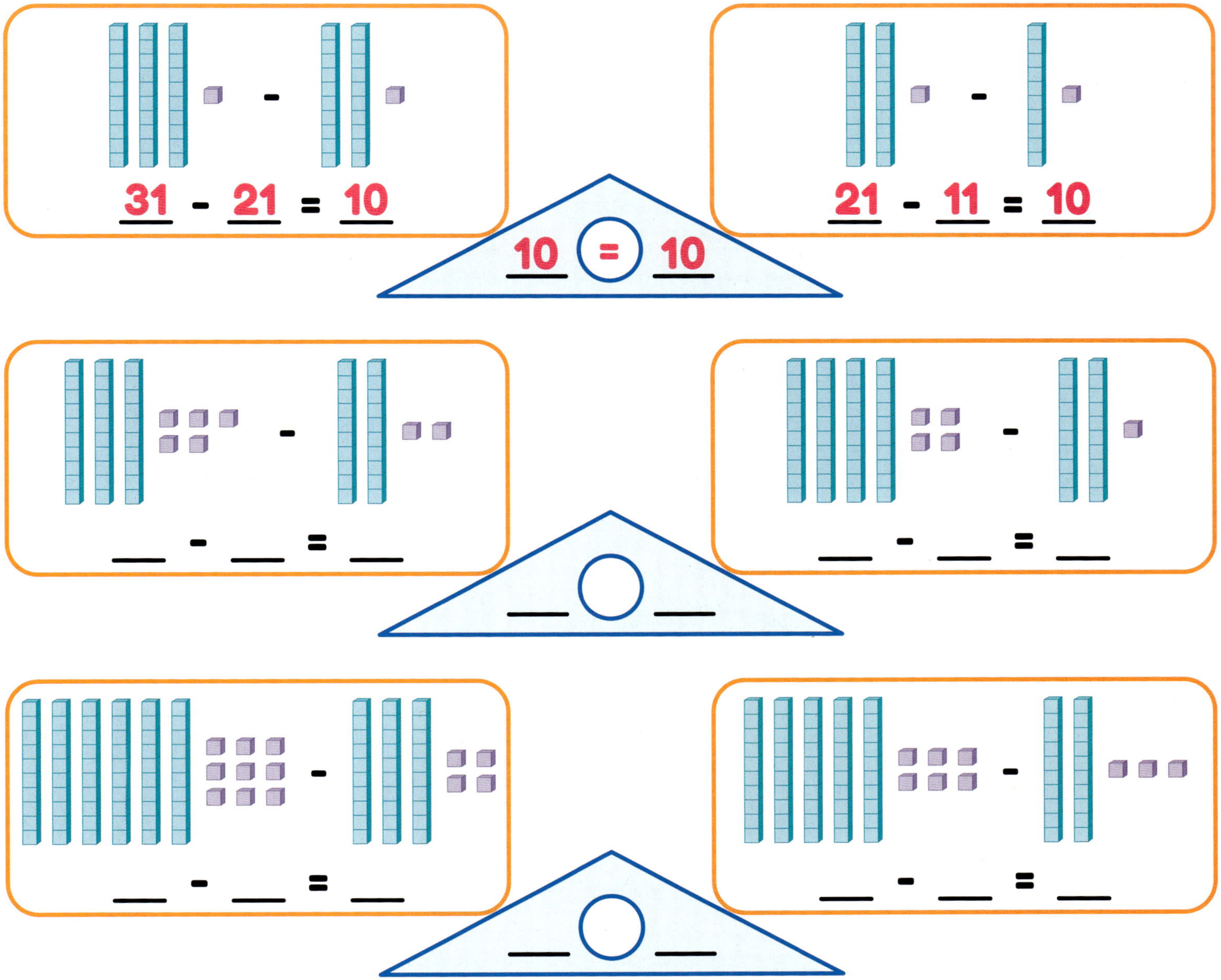

Fun Fruit

A Fruitful Day

Add some colorful fruit to your day. In each place value problem, write the sum in the shaded area. Then, use the same color to color the fruit with the matching number.

Tens	Ones
3	0
+ 3	1

Tens	Ones
4	6
+ 1	3

Tens	Ones
5	8
+ 1	0

Tens	Ones
2	5
+ 7	4

Tens	Ones
6	2
+ 3	6

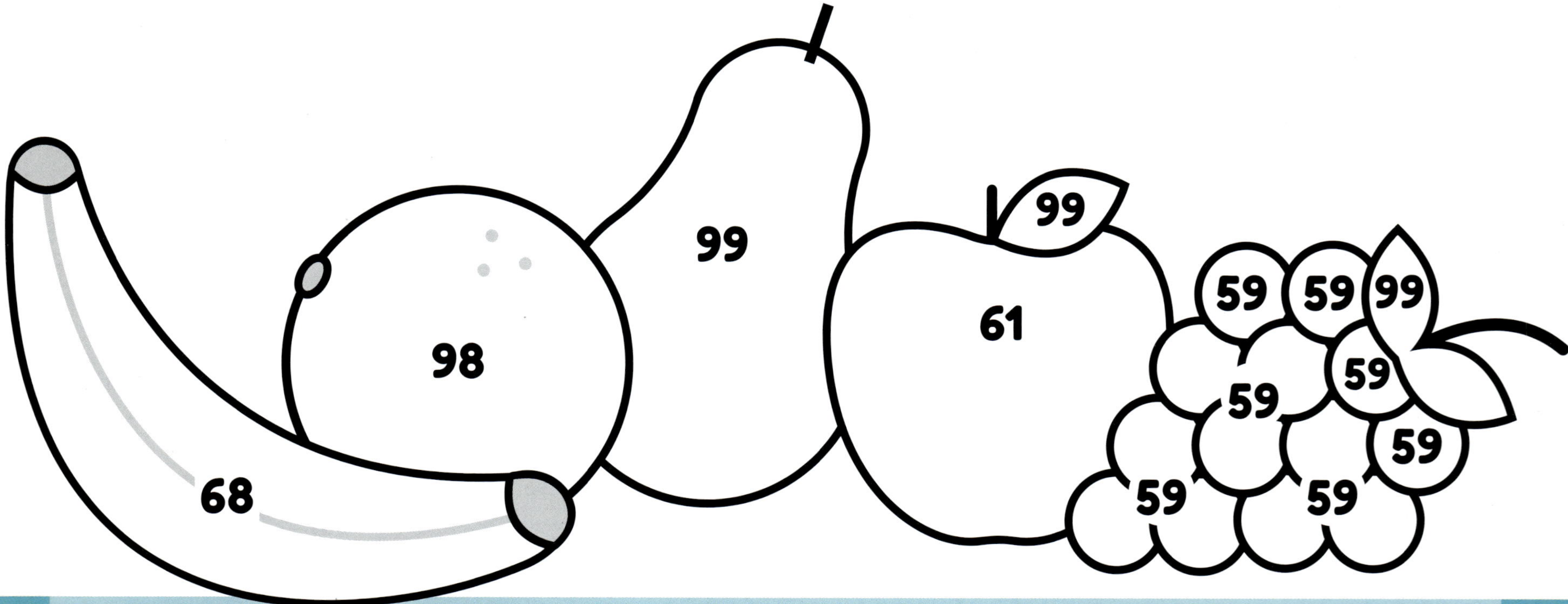

Materials needed: crayons or colored pencils in red, purple, yellow, green, and orange

Land Ho!

Can you follow the map to the island?

- Begin at Start.
- Look at the problem and move along the path that shows the correct answer.
- Continue solving the problems to follow the path of correct answers until you reach the island.

Start

Read

The Collector

Your child is ready to read *The Collector*. As they learn about the narrator's fascination with gathering groups of objects, they are also exposed to many examples of two-digit addition and subtraction.

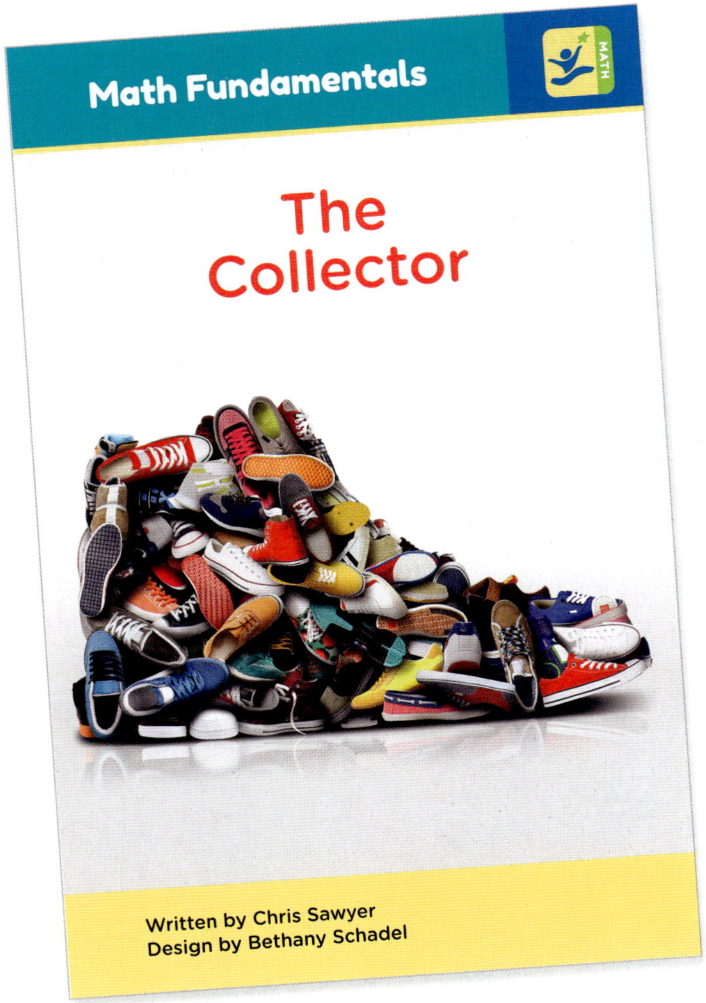

During reading:

Look at the different objects. Ask your child if they can think of other ways to sort the collection.

Encourage your child to find the bonus items in each collection.

After reading:

Does your child have any collections of their own? If so, encourage them to count the number of objects in their collection.

Which of the book's collections did your child like best? What would your child like to collect?

Have your child count their socks. How many pairs of socks do they own all together? How many pairs are white? How many pairs are colored? Using those numbers, invite your child to create and solve a two-digit addition or subtraction problem like those in the book.

Celebrate your child's progress by placing a star sticker in the workbook.

Let's Go Shopping!

Spin Spinner 1 and write the number in the top box. Spin Spinner 2 and write the number in the box below it. Add up the two numbers and write the sum. Place a check next to the item that has the same sum. Repeat for all the empty boxes.

+	+	+	+	+

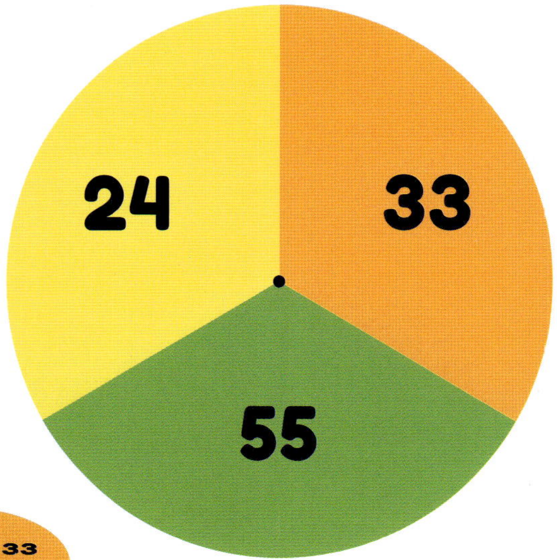

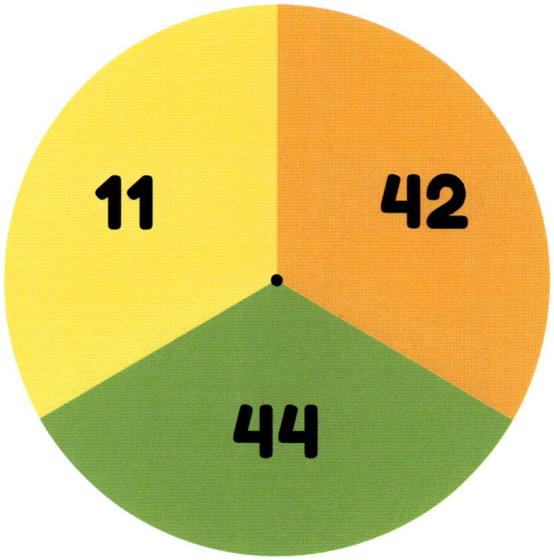

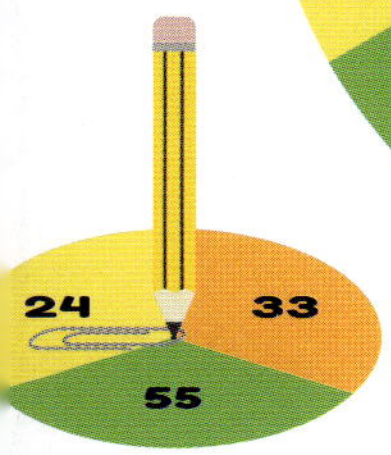

Check		
☐	35	
☐	44	
☐	66	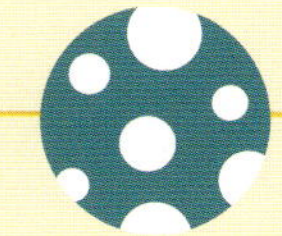
☐	68	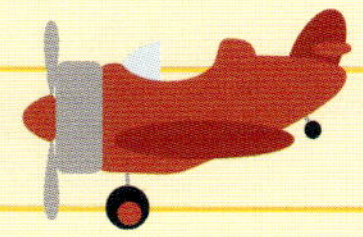
☐	75	
☐	77	
☐	97	
☐	99	

Your child has completed Step 11!
Celebrate by placing a star sticker in the workbook.

HOOKED ON MATH

It's Time to Celebrate!

has completed Math Fundamentals Steps 10 & 11
and is now ready to move on to the
ELEMENTARY MATH level!

SKILLS MASTERED

- Estimation
- Representing numbers
- Place value blocks to 120
- Comparing numbers to 120
- Addition and subtraction with place value blocks
- Two-digit addition and subtraction

Don't forget to share your success and tag us on social media so we can celebrate with you!

Credits

Design Team
Julie Linden Frazier
Lindsey Frazier

Editorial Team
Donna Giachetti
Cynthia Stierle
Austin Feld

Illustrations
Dubit Limited

Educational Advisor
Lesley Freed, Curriculum Specialist

Published by Sandviks HOP, Inc.

Printed in China

HOOKED ON MATH

Math Fundamentals

18031 (07/22)